PROGRESS IN COLLOID & POLYMER SCIENCE

Editors: H.-G. Kilian (Ulm) and G. Lagaly (Kiel)

Volume 91 (1993)

Application of Scattering Methods to the Dynamics of Polymer Systems

Guest Editors:
B. Ewen (Mainz), E. W. Fischer (Mainz),
and G. Fytas (Heraklion)

Springer-Verlag
Berlin Heidelberg GmbH

ISBN 978-3-662-15694-0   ISBN 978-3-7985-1678-6 (eBook)
DOI 10.1007/978-3-7985-1678-6
ISSN 0340-255 X

© 1993 by Springer-Verlag Berlin Heidelberg
Originally published by Dr. Dietrich Steinkopff Verlag GmbH & Co. KG, Darmstadt in 1993
Softcover reprint of the hardcover 1st edition 1993

Chemistry editor: Dr. Maria Magdalene Nabbe; Production: Holger Frey, Thomas Broll

Type-Setting: Macmillan Ltd., Bangalore, India

Preface

The 27th Europhysics Conference on Macromolecular Physics was held in Crete, Greece from September 23–27, 1991. It was organized jointly by the Foundation for Research and Technology Hellas (FORTH) and the Max-Planck-Institut für Polymerforschung in Mainz under the sponsorship of the European Physical Society.

The meeting focused on applications of scattering methods to the dynamics of polymer dense systems and covered Rayleigh–Brillouin scattering and photon correlation spectroscopy, quasi-elastic neutron scattering, holographic methods, real time X-ray and neutron scattering techniques as well as the treatment of theoretical models and computer simulations of polymer dynamics. The present issue contains concise papers presented at the meeting after they have been reviewed.

The meeting was attended by over 100 participants from 16 countries. The scenery of the Agia Pelagia peninsula not far from the Minoan Palace, provided a pleasant setting for stimulating, fruitful discussions besides recreation.

We wish to thank: The members of the scientific committee for advice in the preparation of the program, the sponsors (FORTH, Max-Planck-Gesellschaft zur Förderung der Wissenschaften e.V., Ministry of Science and Technology and the companies A.G. Petzetakis, Plastica Kritis, AMCO) as well as the participants, the authors and the referees of the submitted papers.

B. Ewen
E. W. Fischer
G. Fytas

Contents

Preface. V

Bahar I, Erman B, Kremer F, Fischer EW: Dynamic rotational isomeric state approach for segmental motions of *cis*-polyisoprene in the bulk state 1

Binder K, Paul W, Wittmann H-P, Baschnagel J, Kremer K, Heermann DW: Computer simulation of the glass transition of polymer melts 5

Brereton MG: The dynamics of polymer blends: interdiffusion and the glass transition 8

Darinski A, Gotlib Yu, Lukyanov M, Lyulin A, Neelov I: Computer simulation of the molecular motion in LC and oriented polymers 13

Erman B, Bahar I: Local dynamics of freely rotating polymer chains in dense systems 16

Alvarez F, Colmenero J, Wang CH, Fytas G: Relaxation behaviour in bulk PIMA and PIMA-PMMA copolymer near T_g. 20

Arbe A, Alegría A, Alvarez F, Colmenero J, Frick B: Dynamics of the α-relaxation in glass-forming polymeric systems. Study by neutron scattering and relaxation techniques 24

Floudas G, Higgins JS, Burgess A: Incoherent quasielastic neutron scattering study of a glass-forming liquid. A mode coupling interpretation. 28

Lohfink M, Sillescu H: Tracer diffusion in polymer and organic liquids close to the glass transition studied by forced Rayleigh scattering. 31

Patkowski A, Fischer EW, Gläser H, Meier G, Nilgens H, Steffen W: Light scattering studies of a glass forming liquid near T_g 35

Schönhals A, Kremer F, Schlosser E: The scaling of the α-relaxation in polymers and low-molecular glass-forming liquids – a comparison 39

Sidebottom DL, Bergman R, Börjesson L, Torell LM: Scaling behaviour in poly(propylene glycol) in the glass transition range 43

Torell LM, Jacobsson P, Sidebottom D, Börjesson L: Structural relaxation characteristics of glass-forming polymeric liquids subject to transient cross-links 46

Chu B, Li Y: Synchrotron SAXY studies of segmented polyurethanes 51

Fleischer G: Investigations of self- and tracer diffusion in poly(ethylene oxides) and in blends of poly(dimethyl/ethylmethyl siloxanes) with the pulsed field gradient NMR. 55

Gerharz B, Vogt S, Fischer EW, Fytas G: Segmental relaxation in a symmetric poly(styrene-b-methylphenylsiloxane) copolymer in the disordered phase 58

Hoffmann A, Koch T, Schuler M, Stickel F, Stühn B: Static and dynamic scattering at the microphase separation transition in block copolymers. 61

Meier G, Momper B, Fischer EW: Mode coupling corrections to the Onsager coefficient as determined by light scattering of critical concentration fluctuations from polymer mixtures. 66

Mortensen K: PEO-PPO-PEO block polymer in aqueous solution: Micelle formation and crystallization 69

Rizos AK, Fytas G, Roovers JEL, Ngai KL: Dynamic light scattering study of a 1,4-isoprene-b-styrene copolymer 72

Roland CM, Ngai KL: Concentration fluctuations and segmental relaxation in miscible polymer blends 75

Schwahn D, Janßen S, Springer T: Exponential and non-exponential relaxation and early state of spinodal decomposition in polymer blends by SANS 80

Toprakcioglu C, Dai L, Ansarifar MA, Stamm M, Motschmann H: Equilibrium and dynamic aspects of end-attached diblock and triblock copolymer chains. 83

Anastasiadis SH, Menelle A, Russell TP, Satija SK, Majkrzak CF: Very thin films of symmetric diblock copolymers 88

Reiter G, Steiner U: Short-time dynamics of polymer diffusion across an interface. 93

Russell TP, Menelle A, Anastasiadis SH, Satija SK, Majkrzak CF: The ordering of thin films of symmetric diblock copolymers 97

Stamm M, Götzelmann A, Gießler KH, Rauch F: Organized structures in diblock copolymer films of polystyrene and poly-para-methylstyrene 101

Bastide J, Boué F, Mendes E, Zielinski F, Buzier M, Lartigue C, Oeser R, Lindner P: Is the distribution of entanglements homogeneous in polymer melts? 105

Benmouna M, Fischer EW, Benoit H, Benmansour Z, Vilgis TA: Dynamic scattering from ternary mixtures of polymers in solution. 109
Brown W: Dynamics in concentrated polymer solutions studied using dynamic light scattering 113
Duval M, Haida H, Lingelser JP, Gallot Y: Dynamics of PS-PMMA diblock copolymers in toluene 117
Ewen B, Richter D, Farago B, Maschke U: The effect of microscopic spatial restrictions on the segmental diffusion of dense polymer systems: Their observation and analysis by neutron spin echo spectroscopy 121
Floudas G, Steffen W, Giebel L, Fytas G: Polymer and solvent dynamics in a polystyrene/di-2-ethylhexyl phthalate solution 124
Helmstedt M: Dynamic light scattering investigations on semidilute solutions of branched polyethylene. 127
Richter D, Ewen B, Fetters LJ, Huang JS, Farago B: On the dynamic of dense polymer systems 130
Rizos AK, Ngai KL, Fytas G: Solvent reorientation dynamics in Aroclor/polymer solutions 135
Wang CH: Dynamic light scattering and viscoelasticity in polymer solutions 138
Chu B, Yu J, Wang Z: Dynamics of polymer chains from expanded coils to the collapsed state 142
Hellmann GP, Hellmann EH, Rennie AR: Chain fragmentation and fragment diffusion at the glass transition . . 146
Joosten JGH: Dynamic light scattering by non-ergodic media. 149
Langley KH, Teraoka I, Karasz FE: Diffusion of flexible and semirigid polymers confined to the pore spaces in porous glass 153
Ricka J, Meewes M, Quellet Ch, Binkert Th: Coils, globules and solubilization of a thermosensitive polymer . . 156
Schlosser E, Schönhals A: Relation between main- and normal-mode relaxation. A dielectric study on poly(propyleneoxide) 158
Stieber F, Floudas G, Alig I, Fytas G: Structural relaxation in a low molecular weight poly(methylphenyl siloxane) 162
Tracy MA, Pecora R: Diffusion in rod/sphere composite liquids 165

Author Index 171

Subject Index 172

Progress in Colloid & Polymer Science   Progr Colloid Polym Sci 91:1–4 (1993)

Dynamic rotational isomeric state approach for segmental motions of *cis*-polyisoprene in the bulk state

I. Bahar[1]), B. Erman[1]), F. Kremer[2]), and E. W. Fischer[2])

[1]) Polymer Research Center and School of Engineering, Bogazici University, Istanbul, Turkey
[2]) Max-Planck-Institut für Polymerforschung, Mainz, FRG

Abstract: Dynamic rotational isomeric state formalism is used to compute dipolar correlation functions for *cis*-polyisoprene in the bulk state. Transitions between rotational isomeric states take place through coupled motion of triplets of neighboring bonds in a given repeat unit, as follows from the study of intramolecular conformational energetics in the polymer. The intermolecular effect on local chain dynamics is included in two stages. First the relaxation in a homogeneous environment is treated through adoption of a local effective frictional resistance increasing with the size of the kinetic segment. Secondly, free-volume or density fluctuations of the medium are approximated by a bistate environment. The frequency distribution of relaxational modes is found to broaden with increasing number of bonds cooperatively participating in the segmental mode process. From a comparison with recent dielectric measurements of bulk *cis*-polyisoprene, the experimentally observed Kohlrausch–Williams–Watts (KWW) exponent of 0.39 is attributed to the cooperative relaxation of a kinetic segment of about three repeat units in a fluctuating environment.

Key words: Segmental motions – local chain dynamics – *cis*-polyisoprene in the bulk state – dynamic rotational isomeric state formalism – fluctuating environment

Introduction

The time decay of the orientational correlation function $\phi(t)$ for both segmental and normal modes of relaxation processes in polymers is commonly approximated by the empirical Kohlrausch–Williams–Watts (KWW) or the stretched exponential function

$$\phi(t) = \exp\left[-(t/\tau)^\beta\right] , \qquad (1)$$

where τ is a characteristic correlation time and the exponent β is less than unity. Normal modes cover large portions of the polymer chains, and scale with the second power of the molecular weight, whereas segmental motions are associated with local configurational transitions of backbone bonds and depend on the local chemical structure.

Segmental modes in the bulk *cis*-polyisoprene (*cis*-PIP) have been explored by several experimental techniques such as dielectric relaxation [1, 2], time-resolved optical fluorescence [3] and carbon-13 NMR measurements [4]. β for the segmental mode in *cis*-PIP is reported [1] to be 0.39 and is independent of temperature, whereas it is larger than 0.5 for the normal mode and increases with temperature. A model incorporating (i) the system-specific chemical and structural features of local dynamics, based on the dynamic rotational isomeric state formalism [5], and (ii) the environmental contribution associated with local density fluctuations is presented below for the investigation of the segmental mode relaxation process of *cis*-PIP chains in the bulk state, above the glass transition temperature.

Master equation formalism

The sequence of bonds cooperatively contributing to segmental relaxation is referred to as the "kinetic segment". Their orientational motion relative to a laboratory-fixed frame appended to the first bond of the kinetic segment is studied. A set of torsional angles $\{\Phi\}_\alpha = \{\phi_1, \phi_2, \ldots, \phi_n\}$ specifies a given configuration α assumed by the kinetic segment. The subscript α in $\{\Phi\}_\alpha$ varies in the range $1 \leq \alpha \leq N$, where N denotes the total number of configurations. $\mathbf{P}(t)$ represents the N-dimensional vector of the probabilities $P_\alpha(t)$ for all possible configurations $\{\Phi_\alpha\}$. $\mathbf{P}(t)$ obeys the master equation

$$\frac{d\mathbf{P}(t)}{dt} = \mathbf{A}\mathbf{P}(t) , \qquad (2)$$

where $\mathbf{A}$ is the $N \times N$ transition rate matrix for the segment. The element A_{ik} of $\mathbf{A}$ represents the rate constant for the passage from configuration $\{\Phi\}_k$ to $\{\Phi\}_i$. For stationary processes, the diagonal elements of $\mathbf{A}$ are found from $A_{kk} = -\sum_i A_{ik}$. Here the summation is performed over all of the states i which are different from k. Equation (2) may be solved formally to yield

$$\mathbf{P}(t) = \mathbf{B} \exp(\Lambda t) \mathbf{B}^{-1} \mathbf{P}(0) = \mathbf{C}(t)\mathbf{P}(0) , \qquad (3)$$

where $\mathbf{P}(0)$ is the array of the initial probabilities of the N accessible configurations, $\mathbf{B}$ is the matrix of eigenvectors of $\mathbf{A}$, Λ is the diagonal matrix of eigenvalues of $\mathbf{A}$, $\mathbf{B}^{-1}$ is the inverse of $\mathbf{B}$, and $\mathbf{C}(t) \equiv \mathbf{B} \exp(\Lambda t) \mathbf{B}^{-1}$ is the time-delayed conditional probability matrix. Its elements $C_{\alpha\beta}(t)$ describe the conditional probability of the occurrence of configuration $\{\Phi\}_\alpha$ at time t, given the configuration $\{\Phi\}_\beta$ at $t = 0$. The orientational correlation function $M_1(t)$ for any two vectors $\mathbf{m}$ and $\mathbf{n}$ rigidly embedded in the chain is found from the summations performed over all configurations of the segment as

$$M_1(t) \equiv \langle \mathbf{m}(0) \cdot \mathbf{n}(t) \rangle$$
$$= \sum_{\alpha=1}^{N} \sum_{\beta=1}^{N} C_{\alpha\beta}(t) P_\beta(0) \mathbf{m}_\alpha \cdot \mathbf{n}_\beta , \qquad (3)$$

where $\mathbf{m}_\alpha$ and $\mathbf{n}_\beta$ denote the vectors $\mathbf{m}$ and $\mathbf{n}$ in configurations α and β, respectively. In component form, $M_1(t)$ reads

$$M_1(t) = \sum_{\xi=1}^{N} k_\xi \exp\{\lambda_\xi t\} , \qquad (4)$$

where the frequencies $|\lambda_\xi|$ are the eigenvalues of $\mathbf{A}$,

and the amplitude factor k_ξ is given by

$$k_\xi \equiv \sum_{\alpha=1}^{N} \sum_{\beta=1}^{N} [B_{\alpha\xi} B_{\xi\beta}^{-1} P_\beta(t=0) \mathbf{m}_\alpha \cdot \mathbf{n}_\beta] . \qquad (5)$$

Transition rates matrix

Molecular mechanics type analysis of intra-molecular conformational energies as a function of the torsional angles of interdependent backbone bonds indicates that rotational states corresponding to dihedral angles of $0°$, $\pm 120°$ or $\pm 80°$, referred to as trans (t), gauche$^\pm$ (g$^\pm$) or skew$^\pm$ (s$^\pm$), are assumed by backbone bonds, depending upon their location in the repeat unit. Interdependence of neighboring bonds reduces the number of accessible conformations to a total of 12, in a homogeneous medium [6]. Using the indices shown in Fig. 1a for the backbone bonds of repeat units, the

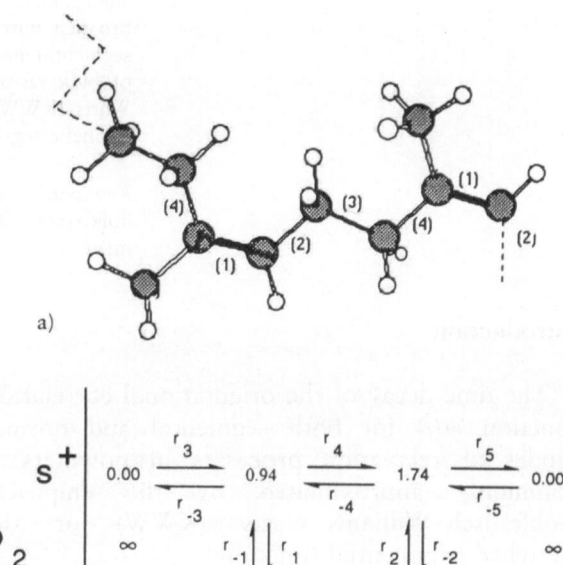

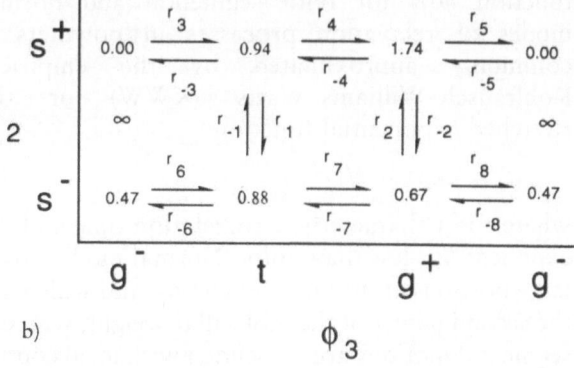

Fig. 1. Portion of the *cis*-polyisoprene chain, indicating bond indices. The double bonds are shown by the bold lines. (b) Kinetic scheme of isomeric transitions accessible to bonds 2 and 3 of a repeat unit, given that bond 4 is in the s$^+$ state

kinetic scheme for the rotameric transitions between half of the 12 states is represented by Fig. 1b. A similar scheme applies to their mirror images. In the case of a bistate fluctuating environment [6] on the other hand, joint states comprising both the state of the chain segment and that of its close neighborhood should be considered, which leads to a value of $N = 24$, for the total number of configurations per repeat unit. Accordingly, the rate matrix A_k for the kth repeat unit of the kinetic segment is organized as

$$A_k = \begin{bmatrix} A_k^* - r_{fast}E_{12} & r_{slow}E_{12} \\ r_{fast}E_{12} & gA_k^* - r_{slow}E_{12} \end{bmatrix}, \quad (6)$$

where E_{12} is the identity matrix of order 12, r_{slow} is the rate of transition of the bistate fluctuating environment from the slow to the fast state. The reverse transition rate r_{fast} is fixed by the detailed balance principle as $r_{slow}/r_{fast} = (1 - p_{slow})/p_{slow}$. Here, p_{slow} is the fraction of the slow state in the medium or the probability of occurrence of a slow environment. The coefficient $g < 1$ accounts for the decreased mobility of the chain in the slow medium, relative to the fast medium. It equals the ratio of the values assumed by the effective friction coefficient $\zeta_{i,eff}$ for bond i in the slow and fast media. A_k^* is the 12×12 transition rate matrix corresponding to the kth monomeric unit with respect to a chain-embedded local coordinate system $Oxyz$, affixed to a double bond. A_k^* operates in a fast homogeneous environment. The explicit form of A_k^* is given in Table IV of ref. [6]. The off-diagonal elements of A_k^* are rate constants given by

$$r_i = \frac{(\gamma^* \gamma)^{1/2}}{2\pi\zeta_{i,eff}} \exp\{-E_{act}/RT\}, \quad (7)$$

where γ and γ^* refer to the curvature of the energy path at the minimum and saddle point, respectively, E_{act} is the activation energy for the specific transition, and $\zeta_{i,eff}$ is the effective friction coefficient for bond i, belonging to the kth repeat unit. Here i refers to serial position with respect to $Oxyz$. For the analysis of the orientational motion of a given dipole moment vector **m** located at the nth bond relative to $Oxyz$, a quadratic dependence of $\zeta_{i,eff}$ on the number of intervening bonds between the dipole moment vector and the rotating bond i is adopted. Thus, $\zeta_{i,eff}$ is conveniently taken

as $\zeta_{i,eff} = A_0(T)/(n-i)^2$, where $A_0(T)$ is a proportionality constant with WLF-type dependence on temperature.

In the more realistic case of x repeat units collectively participating in the relaxation process, the operating transition rate matrix $A^{(x)}$ is found from the direct product of the individual transition matrices A_k with E_{12} as

$$A^{(x)} \equiv \{A_1 \otimes E_{12} \otimes E_{12} \cdots \otimes E_{12}\}$$
$$+ \{E_{12} \otimes A_2 \otimes E_{12} \cdots \otimes E_{12}\} + \cdots$$
$$+ \{E_{12} \otimes E_{12} \otimes E_{12} \cdots \otimes A_x\}. \quad (8)$$

The subscripts $1, 2, \ldots, x$ appended to A distinguish between consecutive transition matrices, as the position-dependent friction coefficient in Eq. (7) implies.

Calculations and comparison with experiments

The activation energies in the rate expressions are evaluated from a detailed quantitative analysis of isomeric minima and saddle heights in conformational energy surfaces and a mathematically efficient matrix multiplication scheme is employed for the evaluation of correlation functions, as the num-

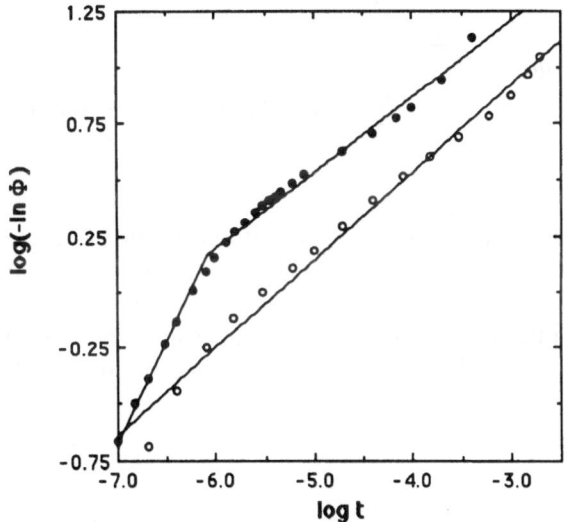

Fig. 2. Dielectric response function for a kinetic segment of three monomer units. Filled and empty circles represent the results for the homogeneous and the bistate fluctuating environments, respectively

ber of transitions contributing to average dynamic properties amount to 24^{2x} for a kinetic segment of x repeat units, in a heterogeneous environment [6].

Calculations are performed for $T = 240$ K, which lies within the narrow temperature range studied in dielectric experiments [1]. The results obtained for a *homogeneous* environment exhibit a stretched exponential decay of orientational correlations at long times, preceded by a single exponential behavior during the initial stages of relaxation. Approximately $1/e$ point of full relaxation appears as the region where the passage from one regime to the other takes place. The exponent β in the long-time regime is found to decrease with the size of the kinetic segment engaged in the cooperative relaxation process. It equates to 0.62, 0.39 and 0.34 for kinetic segments of one, two and three monomer units, respectively. The fact that the KWW-type behavior predicted by the theory is confined to the long-time tail of the correlation decay curves only, in contrast to the experiments, stipulates the necessity of broadening the frequency distribution of relaxational modes. This requirement is achieved by allowing for the density fluctuations of the environment. Figure 2 displays the time dependence of the normalized response function $\Phi(t)$, given by

$A_0 = 1.0 \times 10^9/\text{s}$, $p_{\text{slow}} = 0.5$, $r_{\text{slow}} = 1.0 \times 10^3/\text{s}$, and $g = 1.0 \times 10^{-3}$. The results for a homogeneous environment are also shown by the filled circles, for comparison. The exponents from the slopes of the best-fitting two lines through the filled circles are 0.92 and 0.34, while the empty circles are approximated by a straight line with $\beta = 0.39$, which is in exact agreement with experiments.

References

1. Boese D, Kremer F (1990) Macromolecules 23:829
2. Adachi K, Kotaka T (1984) Macromolecules 17:120; ibid (1985) 18:466; (1987) ibid J Mol Liq 36:75; (1988) Macromolecules 21:157; Imannishi Y, Adachi K, Kotaka T (1988) J Chem Phys 89:7585, 7593
3. Hyde PD, Ediger MD, Kitano T, Ito K (1989) Macromolecules 22:2253
4. Dejean de la Batie R, Laupretre F, Monnerie L (1989) Macromolecules 22:122
5. Bahar I, Erman B (1987) Macromolecules 20:1368
6. Bahar I, Erman B, Kremer F, Fischer EW (1992) Macromolecules 25:816

Received December 17, 1991;
accepted June 17, 1992

$$\Phi(t) = \frac{\sum_{(s)=1}^{x} \sum_{(r)=1}^{x} [\langle \mathbf{m}_{(s)}(t) \cdot \mathbf{m}_{(r)}(0) \rangle - \langle \mathbf{m}_{(s)}(0) \rangle \cdot \langle \mathbf{m}_{(r)}(0) \rangle]}{\sum_{(s)=1}^{x} \sum_{(r)=1}^{x} [\langle \mathbf{m}_{(s)}(0) \cdot \mathbf{m}_{(r)}(0) \rangle - \langle \mathbf{m}_{(s)}(0) \rangle \cdot \langle \mathbf{m}_{(r)}(0) \rangle]}, \tag{9}$$

where $\mathbf{m}_{(s)}(t)$ denotes the dipole moment located at the sth repeat unit of the kinetic segment. Empty circles represent the results for the case of a kinetic segment of three repeat units relaxing in a bistate fluctuating environment, obtained with

Authors' address:

Ivet Bahar
Associate Professor
Polymer Research Center and School of Engineering
Bogazici University, Bebek 80815
Istanbul, Turkey

Progress in Colloid & Polymer Science   Progr Colloid Polym Sci 91:5–7 (1993)

Computer simulation of the glass transition of polymer melts

K. Binder, W. Paul, H.-P. Wittmann, J. Baschnagel, K. Kremer[1]), and D. W. Heermann[2])

Institut für Physik, Johannes Gutenberg-Universität Mainz, FRG
[1]) IFF, Forschungszentrum Jülich, FRG
[2]) Institut für Theoretische Physik, Universität Heidelberg, FRG

Abstract: Bond fluctuation models on square and simple cubic lattices at melt densities are simulated, using potentials depending on the length of the (effective) bond (and also on the bond angle, in $d = 3$ dimensions). Various relaxation functions have the Kohlrausch–Williams–Watts (KWW) form; the associated relaxation time diverges as $\exp(\text{const}/T^2)$ in $d = 2$ and as $\exp[\text{const}/(T - T_0)]$ in $d = 3$. For $d = 3$ the self-diffusion constant also follows the Vogel–Fulcher law, with $T_0 = 250$ K for chain lengths $N = 20$ and potentials adapted to bisphenol-A-polycarbonate [BPA-PC].

Key words: Glass transition – Monte Carlo method – bond fluctuation model – self-diffusion

Introduction and motivation of the model

We here review recent Monte Carlo studies [1–3] of the thermally driven glass transition of lattice models for polymer melts. The model is the bond fluctuation model [4–7] augmented, in $d = 2$ [1], by a potential for the bond lengths, $U_{\text{eff}}(l) = U_0(l - l_0)^2$. In $d = 3$ [2, 3], in addition a potential $V_{\text{eff}}(\theta) = V_0(\cos\theta - \cos\theta_0)^2$ for the angle between subsequent bonds is used. The motivation for this model is (i) It takes excluded volume into account, and that chains cannot intersect during their motions if the set of allowed bond lengths is suitably restricted ($2 \leq l \leq 14$ in $d = 2$, $2 \leq l \leq \sqrt{10}$ in $d = 3$, $l = \sqrt{8}$ being excluded). (ii) Both in $d = 2$ and in $d = 3$ the single-chain dynamics is Rouse-like [8], as desired. For long chains in $d = 3$ at volume fractions ϕ of occupied sites corresponding to dense melts ($\phi \gtrsim 0.5$ [6, 7], crossover from Rouse dynamics to reptation [8] occurs [7]. At $\phi = 0.5$, the excluded volume screening length is only about two bond lengths [7]. (iii) While the model of [4, 5, 7] is athermal, one can add intermolecular energies [6] or potentials $U_{\text{eff}}(l)$ [1] or $V_{\text{eff}}(\theta)$ [9] or both [2]. Such potentials are motivated by a coarse-graining: n chemical bonds along the backbone of a chain are integrated into

one effective bond [2, 10, 11]. The potentials controlling the lengths of the chemical bond, the bond angles and torsional angles "translate" into effective potentials $U_{\text{eff}}(l)$, $V_{\text{eff}}(\theta)$, mapping the corresponding distribution functions onto each other [2, 10, 11]. The existence of an energetically preferred value l_0 for the "effective bond length" l creates a conflict between configurational entropy (where it is favorable to have many different bond lengths l at disposal) and energy and, thus, is ultimately responsible for having a glass transition in this model. (iv) Very efficient simulation codes exist, with a speed of 2–$3 \cdot 10^6$ attempted monomer jumps per second, and a similar performance on parallel computers. This efficiency allows to study relaxation over large time scales (up to $3 \cdot 10^7$ attempted moves per monomer [7], which means about 10^{-6} s in real time), with high precision ($128\,000$ {$132\,000$} monomers are averaged over in $d = 2$ {$d = 3$}). Such precise data over a wide time range are crucial for a meaningful analysis.

Simulations in two dimensions

Choosing $l_0 = \sqrt{10}$, 20% of all lattice sites cannot be occupied in the ground state [1]. For $\phi \to 1$, a large configurational entropy requires tightly packed monomers, and many bonds have $l = 2$; this choice of l_0 obviously implies a conflict between bond energetics and the tendency towards close packing. Qualitatively, one attributes the glass transition to this "frustration effect" in the model [1]. Figure 1 summarizes findings which resulted from cooling the polymer melt from $1/T = 0$ to a final temperature T_f as $1/T(t) = (1/T_f)\Gamma t$, with $\Gamma = 4 \cdot 10^{-7}$ (times t measured in attempted moves per monomer); U_0

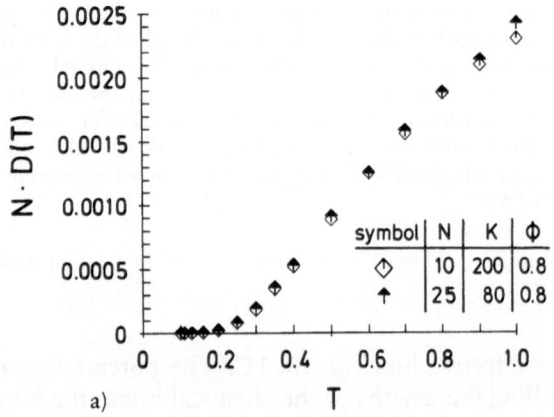

a)

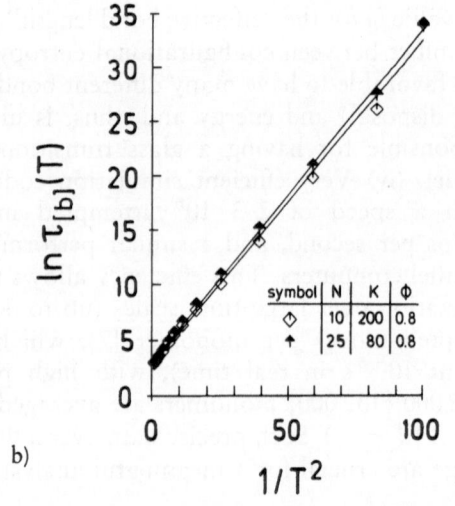

b)

Fig. 1. (a) Normalized self-diffusion constant $ND(T)$ plotted vs. T for $\phi = 0.8$. Note that 64 lattices of size 100×100 with periodic boundary conditions were run in parallel, each lattice containing K chains as shown. (b) Logarithm of the bond orientation autocorrelation time $\tau_{[b]}(T)$ plotted vs. $1/T^2$ (from Wittmann et al. [1])

in the potential $U_{eff}(l) = U_0(l - l_0)^2$ is fixed by $U_{eff}(2) = 1$ (note also $k_B \equiv 1$). Near $T_g \approx 0.2$, the self-diffusion constant practically vanishes (for $T \leq 0.3$ we can no longer estimate $D(T)$, but only extract an upper bound). But near T_g the acceptance rate is clearly nonzero: local motions are still possible. The mobile monomers are "arrested in cages" formed by their environment. The significance of $T_g \approx 0.2$ also shows up by (rounded) kinks in the temperature dependence of compressibility, free volume, etc. [1]. But the relaxation times such as $\tau_{[b]}$ (Fig. 1b) diverge not at T_g but for $T \to 0$ only: at T_g the system "freezes" by falling out of equilibrium, for the chosen cooling rate. This interpretation is supported by the energy which for $T < T_g$ remains frozen rather than reaching the ground state [1]. Probably a static glass transition in $D = 2$ occurs for $T \to 0$ only while at T_g a kinetic freezing is seen.

Simulations in three dimensions

The chosen potentials are $U_{eff}(l)/T_f = a(l - l_0)^2$, $V_{eff}(l)/T_f = b(\cos\theta - \cos\theta_0)^2$, with $a = 4$, $b = 5$, $\theta_0 = 90°$, $l_0 = 3$ for $T_f = 570$ K. These parameters are optimized for BPA-PC, where three effective bonds of the model translate into one monomer [2, 3]; (there are 12 chemical bonds along the backbone within one monomer, which also has much internal flexibility [10].) For $\phi = 0.5$, a density $\rho = 1.05 \, \text{g/cm}^3$ and monomeric mass $M = 425 \cdot 10^{-24}$ g yields a lattice spacing corresponding to 2.03 Å. Indeed, a qualitatively reasonable description of the structure factor $S(Q)$ of the supercooled melt is obtained (Fig. 2). $S(Q)$ is small for small Q, due to the small compressibility. The "amorphous halo" occurs at $Q \approx 1.5 \, \text{Å}^{-1}$ and a weaker peak at larger Q, similar to experiment [12].

Of course, since our calculated $S(Q)$ contains only scattering from pointlike scattering centers at the centers of gravity of the effective monomers, rather than scattering from the individual atoms contained in the real monomers, a quantitatively accurate description of $S(Q)$ for real materials cannot be expected on the present coarse-grained level. For an accurate modelling of $S(Q)$ for BPA-PC (or other amorphous polymers) information on the chemical structure of the monomers needs to be reintroduced into the model.

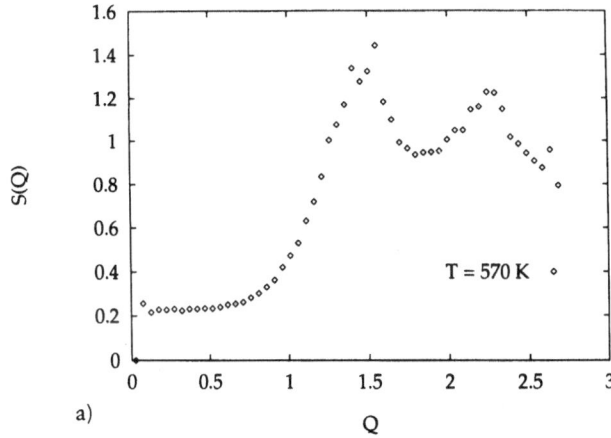

a)

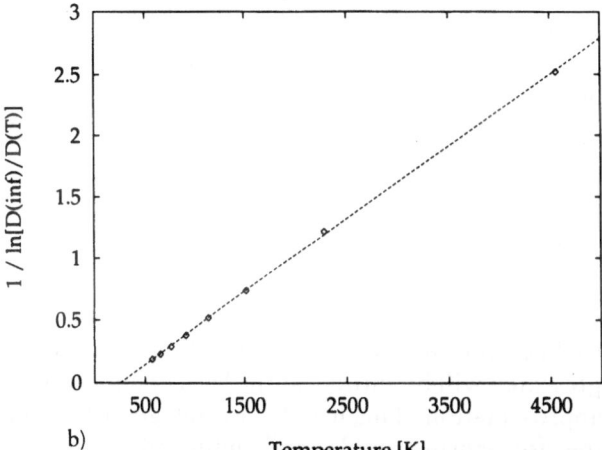

b)

Fig. 2. (a) Static coherent structure factor $S(Q)$ of the lattice model polymer melt resembling BPA-PC at $T = 570$ K plotted vs. Q. (b) Plot of $1/\ln(D_0/D_\infty)$ vs. T for $N = 20$. D_∞ is the diffusion constant for $T \to \infty$. Straight line indicates the Vogel–Fulcher fit, intercept is the Vogel–Fulcher temperature T_0 (from Paul et al. [2, 3])

The self-diffusion is described nicely by the Vogel–Fulcher law, $D_0/D_\infty \sim \exp[-A/(T - T_0)]$ (Fig. 2b), with $A = 1726k_B$, $T_0 = 250$ K. The excitation barrier A is similar to experiment [13]; T_0 is somewhat too low, but this may be due to the smallness of N ($N = 20$ corresponds to only 7 BPA-PC monomers).

Even if this agreement with experiment in Fig. 2b were fortuitous, the model is simple and suitable for a study of glass transitions in general. Many features of real materials are accounted for (e.g. relaxation functions follow the stretched exponential, time–temperature superposition works [3]).

Acknowledgements

Support from the BMFT (grant no. 03M4028) and the BAYER AG, from the DFG (SFB 262/D2) and the HLRZ Jülich is gratefully acknolwedged.

References

1. Wittmann H-P, Kremer K, Binder K (1991) J Chem Phys 96:6291
2. Paul W, Binder K, Kremer K, Heermann DW (1991) Macromolecules 24:6332
3. Paul W, preprint
4. Carmesin I, Kremer K (1988) Macromolecules 21:2819; (1990) J Phys (Paris) 51:915
5. Wittmann II-P, Kremer K (1990) Comp Phys Commun 61:309
6. Deutsch II-P, Binder K (1991) J Chem Phys 94:2294
7. Paul W, Binder K, Heermann DW, Kremer K (1991) J Phys H (Paris) 1:37; J Chem Phys 95:7726
8. Doi M, Edwards SF (1986) Theory of Polymer Dynamics. Clarendon Press, Oxford
9. Lopez-Rodriguez A, Wittmann H-P, Binder K (1990) Macromolecules 23:4327
10. Baschnagel J, Binder K, Paul W, Laso M, Suter UW, Batoulis I, Jilge W, Bürger T (1991) J Chem Phys 95:6014
11. Baschnagel J, Qin K, Paul W, Binder K (1992) Macromolecules 25:3117
12. Richter D, Frick B, Farago A (1988) Phys Rev Lett 61:2465
13. Macho E, Alegria A, Colmenero J (1987) Polym Eng Sci 27:810

Received January 14, 1992; accepted June 13, 1992

Authors' address:

Prof. Dr. K. Binder
Institut für Physik
Johannes Gutenberg-Universität Mainz
Staudingerweg 7
D-W-6500 Mainz, FRG

Progress in Colloid & Polymer Science   Progr Colloid Polym Sci 91:8–12 (1993)

The dynamics of polymer blends: interdiffusion and the glass transition

M. G. Brereton

IRC in Polymer Science and Technology, University of Leeds, UK

Abstract: In this paper it is shown, on the basis of a simple mean field approach to the dynamics of concentration and density fluctuations, that the interdiffusion coefficient $D_{int}(\varphi)$ in a blend (A/B) of composition φ can be related to the self-diffusion coefficients $D_{A/B}^*$ of the pure melts. However, the relation is predicted to be a linear combination of the so-called fast and slow modes:

$$D_{int}^{-1} = (1 - U)\{D_{slow}^*\}^{-1} + U\{D_{fast}^*\}^{-1} \, .$$

The parameter $U = U(T)$ is given and is shown to be a measure of the nearness of the glass transition to the experimental temperature T. Far above the glass transition ($U \to 0$) the interdiffusion coefficient reduces to D_{slow}^* and should show the simple linear dependence on concentration according to

$$\{D_{slow}^*\}^{-1} = \{\varphi D_A^*\}^{-1} + \{(1 - \varphi)D_B^*\}^{-1} \, .$$

Key words: Polymer blends – interdiffusion – glass transition

Introduction

This paper is concerned with an "old" problem concerning the possibility of a relationship between the mutual diffusion coefficient in an A/B binary blend and the self-diffusion coefficients of the pure components. A recent work by Binder et al. [1] provides a useful summary and further references to the work on this topic. The problem is that in the pure melt (all A or B) there are only density fluctuations, whilst in the blend there are additional concentration fluctuations which are driven by diffusive processes. The essence of this work will be to show that there is a coupling between concentration and density fluctuations through the single-particle friction coefficients.

For polymer blends, the original treatment given by de Gennes [2] assumed that the concentration fluctuations were driven entirely by a thermodynamic force. He considered a case of almost identical chains, where, in particular, the glass transition did not change with composition or play any role at all in the mutual diffusion of the species. By considering a coupling between the concentration and density fluctuations it is shown in this paper that the much stronger forces ultimately re-sponsible for the glass transition effects are brought to bear on this problem. The discussion to be presented will be entirely at the level of a mean field approximation. This has the advantage of keeping the formalism simple and intuitively physical; however, some of the conclusions, particularily about the glass transition, will only be qualitatively correct.

Mean field dynamics of concentration fluctuations

The Fourier components of the concentration fluctuations in a blend of particles located at $\{r_i^A; r_j^B\}$ are given by

$$\rho_q^A = \Omega^{-1}\sum_j \exp i\mathbf{q}\cdot\mathbf{r}_j^A \tag{1}$$

and similarly for ρ_q^B, where Ω is the volume of the system. The dynamics of $\{\rho_q^A\}$ are given directly from the definition as

$$\frac{d}{dt}\rho_q^A = \Omega^{-1}\sum_j\left(i\mathbf{q}\cdot\frac{d}{dt}\mathbf{r}_j^A\right)\exp i\mathbf{q}\cdot\mathbf{r}_j^A \tag{2}$$

and require a knowledge of the dynamical behaviour of the individual particles. For the moment a simple one-component system will be considered, where the interaction potential between the particles is $V(\mathbf{r}_i - \mathbf{r}_j)$. The force acting on the jth particle can be written in terms of the Fourier components V_q of $V(\mathbf{r}_i - \mathbf{r}_j)$ as

$$\mathbf{F}(\mathbf{r}_j) = -\frac{d}{d\mathbf{r}_j} \sum_{ik} V_k \exp i(\mathbf{r}_i - \mathbf{r}_j)$$

$$= -\sum_k V_k(i\mathbf{k}) \exp i\mathbf{r}_j \rho_{-k} . \quad (3)$$

Assuming that the motion of the particles is dominated by the friction, an equation of motion can be written in the form

$$v\frac{d}{dt}\mathbf{r}_j = \mathbf{F}(\mathbf{r}_j) + \xi_j = \eta\{\mathbf{r}_j\} , \quad (4)$$

where v is the actual friction coefficient of the particle in the particular environment. The heat bath forces ξ_j determined a bare friction coefficient v_0 through

$$\langle \xi_j(t)\xi_i(0) \rangle = kT v_0 \delta(t) . \quad (5)$$

Using Eq. (3) for $\mathbf{F}(\mathbf{r}_j)$, Eq. (2) for the density fluctuations becomes

$$\frac{d}{dt}\rho_q = -\frac{1}{v}\sum_k (\mathbf{k}\cdot\mathbf{q}) V_k \rho_{k+q}\rho_k$$

$$+ \text{ heat bath terms .} \quad (6)$$

On the right-hand side of this equation the sum over k involves quadratic terms in the density fluctuations except for the single term $\rho_0\rho_{-q}$ obtained when $k = -q$. This term is only linear in the fluctuations since ρ_0 is related to the (non-fluctuating) macroscopic density by $\rho_0 = N/\Omega = \bar{\rho}$. The mean field approximation consists of neglecting the quadratically fluctuating terms and only retaining this mean field term, i.e.

$$\frac{d}{dt}\rho_q = -\frac{1}{v}q^2 V_q \bar{\rho}\rho_q + \text{heat bath terms.} \quad (7)$$

This can be readily generalized to a two-component blend to give

$$\frac{d}{dt}\rho_q^A = -q^2\left\{V_q^{AA}\frac{\bar{\rho}^A}{v_A}\rho_q^A + V_q^{AB}\frac{\bar{\rho}^A}{v_A}\rho_q^B\right\}$$

$$+ \text{ heat bath terms ,}$$

$$\frac{d}{dt}\rho_q^B = -q^2\left\{V_q^{AB}\frac{\bar{\rho}^B}{v_B}\rho_q^A + V_q^{BB}\frac{\bar{\rho}^B}{v_B}\rho_q^B\right\}$$

$$+ \text{ heat bath terms .} \quad (8)$$

It is important to note that in these expressions the friction coefficients v^A and v^B refer to a particle in the actual A/B blend at a composition φ, i.e. $v^A \equiv v^A(\varphi)$, etc., where $\varphi = \rho^A/\bar{\rho}$. To proceed further with Eq. (7) the interaction potentials are written in the form

$$V_{AA} = V_0 + \varepsilon_{AA}, \quad V_{AB} = V_0 + \varepsilon_{AB} ,$$
$$V_{BB} = V_0 + \varepsilon_{BB} . \quad (9)$$

In the limit $V_0 \gg \varepsilon_{AA}, \varepsilon_{AB}, \varepsilon_{BB}$ the coupled Eqs. (8) are easily solved for $\rho_q^A(t)$ and $\rho_q^B(t)$. The total density fluctuations $\langle \rho_q(t)\rho_{-q}(0) \rangle$ and the concentration fluctuations $\langle \rho_q^A(t)\rho_{-q}^A(0) \rangle$ can be found and written as

$$\langle \rho_q(t)\rho_{-q}(0) \rangle = \langle |\rho_q|^2 \rangle \exp\{-q^2 D_T t\} ,$$
$$\langle \rho_q^A(t)\rho_{-q}^A(0) \rangle = \langle |\rho_q^A|^2 \rangle \exp\{-q^2 D_{int} t\} , \quad (10)$$

where

$$D_T = V_0\bar{\rho}\left\{\frac{\varphi}{v_A} + \frac{(1-\varphi)}{v_B}\right\} = V_0\bar{\rho}/v_{fast} \quad (11)$$

and the interdiffusion coefficient

$$D_{int} = (\varepsilon_{AA} + \varepsilon_{BB} - 2\varepsilon_{AB})\rho\varphi(1-\varphi)/v_{slow} \quad (12)$$

with

$$v_{slow} = \varphi v_B + (1-\varphi)v_A ,$$
$$v_{fast}^{-1} = \varphi v_A^{-1} + (1-\varphi)v_B^{-1} . \quad (13)$$

The designations "fast" and "slow" are made so that if, for example, the A species is slower, i.e. $v_A \gg v_B$, then the density fluctuations are governed by the friction coefficient of the fast B particles:

$$D_T = V_0\bar{\rho}\frac{(1-\varphi)}{v_B} \quad \text{(fast mode) ,} \quad (14)$$

while the concentration fluctuations are governed by the slow species

$$D_{int} = (\varepsilon_{AA} + \varepsilon_{BB} - 2\varepsilon_{AB})\frac{\rho\varphi}{v_A} \quad \text{(slow mode) .} \quad (15)$$

The mean field approximation predicts that the interdiffusion is controlled by the "slow-mode" result:

$$D_{\text{int}} \sim 1/v_{\text{slow}}(\varphi),$$

where

$$v_{\text{slow}}(\varphi) = \varphi v_{\text{B}}(\varphi) + (1 - \varphi) v_{\text{A}}(\varphi) . \qquad (16)$$

In the original de Gennes treatment the terms corresponding to v_{A} and v_{B} were considered to be independent of the concentration φ and to refer to the pure melts. Hence, the concentration dependence of the interdiffusion coefficient D_{int} was predicted to be the linear behaviour contained in v_{slow}. However, other workers [3, 4] predicted that D_{int} should involve the friction term v_{fast} rather than v_{slow}. Experimentally, the matter does not seem so straightforward to decide. Evidence for the slow-mode result has been presented by [5, 6] and for the fast mode by [7–9]. The result (Eq. (16)) emphasizes that the φ dependence of v is not linear, since v_{A} and v_{B} relate to the actual blend at the composition φ and not to any limiting values associated with the pure states.

In the next section the problem of finding an expression for the friction coefficients v_{A} and v_{B} will be considered. In general, friction coefficients are closely related to the time integrals over the fluctuating environmental forces. Within the mean field approximation it will be shown that the fluctuating forces in turn depend on the density and concentration fluctuations present in the system. These later quantities have already been determined in Eq. (10) and depend on the friction coefficients that are to be determined through Eqs. (11)–(13). Simple self-consistent equations are derived for the friction coefficients, which also show a (mean field) glass transition.

A mean field glass transition and the friction coefficient

As before, consider a one-component system with a dynamics described by Eq. (4). The friction coefficient is determined by the fluctuations of the "environmental force" $\eta\{r_j\}$ through the fluctuation dissipation result

$$v = \int_0^\infty dt \langle \eta\{r_j(t)\} \eta\{r_j(0)\} \rangle / kT$$

$$= v_0 + \int_0^\infty dt \langle F\{r_j(t)\} F\{r_j(0)\} \rangle / kT . \qquad (17)$$

Using expression (3) for $F(r_j)$,

$$\langle F\{r_j(t)\} F\{r_j(0)\} \rangle = \sum_{kq} -k \cdot q V_k V_q$$

$$\times \langle \exp ik \cdot r_j(t) - iq \cdot r_j(0)) \rho_k(t) \rho_q(0) \rangle . \qquad (18)$$

The single-particle terms are decoupled from the density fluctuations in the spirit of the mean field approximation to give

$$\langle \exp ik \cdot r_j(t) - iq \cdot r_j(0) \, \rho_k(t) \rho_q(0) \rangle$$

$$\approx \langle \exp ik(r_j(t) - r_j(0)) \rangle \langle \rho_k(t) \rho_{-k}(0) \rangle .$$

The density fluctuations $\langle \rho_k(t) \rho_{-k}(0) \rangle$ have already been found in Eq. (10) and are dependent on the friction coefficient v through D_T given by Eq. (11). It is also assumed that they are much faster than the single-particle term so that the time integral gives a term directly proportional to v:

$$\int_0^\infty dt \langle \rho_k(t) \rho_{-k}(0) \rangle \langle \exp ik[r_j(t) - r_j(0)] \rangle$$

$$\approx \int_0^\infty dt \langle \rho_k(t) \rho_{-k}(0) \rangle$$

$$= \langle |\rho_k|^2 \rangle \int_0^\infty dt \exp \left[\frac{-k^2 V_0 \bar{\rho} t}{v} \right]$$

$$= v \langle |\rho_k|^2 \rangle / \{k^2 V_0 \bar{\rho}\} . \qquad (19)$$

Finally, Eq. (18) gives a self-consistent equation for the friction coefficient v as

$$v = v_0 + v \sum_k V_0 \langle |\rho_k|^2 \rangle / \bar{\rho} kT$$

or

$$v = v^0 / \{1 - U(T)\} , \qquad (20)$$

where

$$U(T) = \sum_k \frac{V_0 \langle |\rho_k|^2 \rangle}{\bar{\rho} kT} .$$

The result (20) predicts a dynamic mean field "glass transition" at a temperature T_0 when $U(T_0) = 1$. For $T \sim T_0$ this is quantitatively wrong as the friction coefficient diverges not as a power of $(T - T_0)^{-1}$ but according to the Vogel–Fulcher form of $\exp(C/(T - T_0))$, where C is a constant. However, in this paper it is not the intention to discuss the behaviour of the diffusion constants in the neighbourhood of the glass transition. The result (20) will only be used for temperatures $T \gg T_0$, where $U(T) \ll 1$, to indicate the increase of friction as the glass transition is approached, i.e.

$$v = v_0 [1 + U(T)] . \tag{21}$$

The same method, based on Eq. (17), for calculating the friction coefficient can be applied to those appearing in the blend problem. The result is

$$v_A = v_A^0 + (kT)^{-1} \int_0^\infty dt \sum k^2 \{ V_0^2 \langle \rho_k(t) \rho_{-k}(0) \rangle$$
$$+ (\varepsilon_{AA} - \varepsilon_{AB})^2 \langle \rho_k^A(t) \rho_{-k}^A(0) \rangle \} . \tag{22}$$

In this result both the density and concentration fluctuations contribute. Except near a phase separation point the density term will dominate the concentration term since $V_0 \gg (\varepsilon_{AA} - \varepsilon_{AB})$. For the blend the density fluctuations are controlled by the fast-mode combination [Eq. (13)] of friction coefficients v_{fast}. The time integral over the density fluctuations is proportional to v_{fast} and the friction coefficient for an A and B particle in an A/B blend can be written in a form similar to Eq. (20) as

$$v_A = v_A^0 + U(T) v_{fast}$$

or

$$v_B = v_B^0 + U(T) v_{fast} . \tag{23}$$

Finally, if the slow-mode combination $v_{slow} = \varphi v_B + (1 - \varphi) v_A$ of friction coefficients is formed then the major result of this paper can be written as

$$v_{slow} = v_{slow}^0 + U v_{fast} . \tag{24}$$

The theory only has validity for $U(T) \ll 1$, in which case v_{fast} can be replaced by v_{fast}^0 and Eq. (24) becomes

$$v_{slow} = v_{slow}^0 + U v_{fast}^0 . \tag{25}$$

The bare friction coefficients v_A^0 and v_B^0 contained in v_{slow}^0 can be related to the friction coefficients v_A^* and v_B^* of particles in the pure melt by

$$v_A(\varphi = 1) \equiv v_A^* = v_A^0/(1 - U),$$

i.e.

$$v_A^0 = v_A^*(1 - U) ,$$

and, similarly,

$$v_B^0 = v_B^*(1 - U) . \tag{26}$$

Then

$$v_{slow} = (1 - U) v_{slow}^* + U v_{fast}^* . \tag{27}$$

Diffusion coefficients are inversely proportional to the friction coefficients; hence, in terms of the pure melt self-diffusion coefficients D_A^*, D_B^*, the interdiffusion coefficient D_{int} for the blend can be written as

$$D_{int}^{-1} = (1 - U) \{ D_{slow}^* \}^{-1} + U \{ D_{fast}^* \}^{-1} , \tag{28}$$

where

$$D_{fast}^* = (1 - \varphi) D_A^* + \varphi D_B^*$$

and

$$\{ D_{slow}^* \}^{-1} = \{ \varphi D_A^* \}^{-1} + \{ (1 - \varphi) D_B^* \}^{-1} .$$

Conclusions

The result (28) indicates that the interdiffusion coefficient $D_{int}(\varphi)$ in a blend can be related to the self-diffusion coefficients $D_{A/B}^*$ of the pure melts. However, the relation is a linear combination of the slow and fast modes depending on the parameter $U(T)$, which is a measure of the nearness of the glass transition to the experimental temperature T. Far above the glass transition ($U \to 0$), the interdiffusion coefficient reduces to D_{slow}^* and should show the simple linear dependence on concentration associated with $\{ D_{slow}^* \}^{-1}$. This tendency is in fact shown in the computer simulation results of Binder et al. [10]. The simulation is done on a simple lattice and as the number of vacancies is increased the computed interdiffusion constant D_{int} approaches closer to that given by D_{slow}^*. The increased number of vacancies is interpreted in this context as a lowering of the glass transition. The reported [5] experimental verification of the slow-mode result was also achieved by extrapolating the experimental results to high temperatures before testing the fast- and slow-mode theories.

Acknowledgement

This work was done at the Max Planck Institut fur Polymerforschung (Mainz). I would like to thank Prof. E. W. Fischer and Dr. T. Vilgis for the continuing hospitality and useful discussions. I am also grateful to Prof. Binder for his useful comments.

References

1. Binder K, Jilge W, Carmesin I, Kremer K (1990) Macromols 23:5001–5013

2. de Gennes PG (1980) J Chem Phys 72:4756–4763
3. Kramer E, Green P, Palmstrom C (1984) Polymer 25:473–480
4. Sillescu H (1984) Markromol Chem Rapid Commun 5:519
5. Brereton MG, Fischer EW, Fytas G, Murschall U (1987) J Chem Phys 86:5174–5180
6. Fytas G (1987) Macromols 20:1430–1431
7. Jones R, Klein J, Donald AM (1986) Nature 321:16–17
8. Composto R, Mayer J, Kramer E, White D (1986) Phys Rev Letts 57:1312–1315
9. Jordan A, Ball R, Donald AM, Fetters L, Jones A, Klein J (1988) Macromols 21:235–239

10. Binder K, Sariban A, Deutsch H-P (1991) J Non Crystalline Solids 131:635–642

Received November 18, 1991;
accepted May 21, 1992

Author's address:

Dr. M. G. Brereton
Dept. Physics
IRC in Polymer Science and Technology
University of Leeds
Leeds LS2 9JT, UK

Progress in Colloid & Polymer Science   Progr Colloid Polym Sci 91:13–15 (1993)

Computer simulation of the molecular motion in LC and oriented polymers

A. Darinskii, Yu. Gotlib, M. Lukyanov, A. Lyulin, and I. Neelov

Institute of Macromolecular Compounds, St. Petersburg, Russia

Abstract: A computer simulation of the dynamic properties of polymer chains in the presence of a strong nematic field has been carried out by molecular and Brownian dynamics methods. A multichain system, a freely joint chain with rigid bonds and a chain with fixed bond angles and rigid side groups have been studied. The influence of the ordering on the chain conformations, orientational and translational mobility and spectra of relaxation times have been investigated.

Key words: Computer simulation – LC polymers – mobility

A computer simulation of the molecular motion in polymer systems in the presence of a strong quadrupole field

$$U = - U_0 \sum \cos^2 \theta_i \qquad (1)$$

has been carried out. Here θ_i is the angle formed by the long axis of the mesogen with the direction of the field and U_0 is the field magnitude. In some cases the magnetic or electric fields can affect the macromolecules as a quadrupole one. The most important case is the LC polymers, where Eq. (1) is the simplest form of the nematic Maier–Saupe potential, induced by the anisotropic intermolecular interactions. The magnitude U_0 is a function of the order parameter S:

$$S = \left\langle \frac{3}{2} \cos^2 \theta - \frac{1}{3} \right\rangle . \qquad (2)$$

The multichain system of short chains in the external field (1) was studied by the method of molecular dynamics [1]. Every chain consists of 6 Lennard–Jones particles connected by rigid bonds. To study the effect of an oriented environment on the selected chain we have considered a system where the field orients only one chain and the other chains are left undisturbed.

The value of the order parameter S was larger in the system where all chains were under the influence of the field. It means that the oriented chains

affect each other by the additional internal orienting field induced by intermolecular interaction. Suggesting the quadrupole symmetry of this field, the dependence of the magnitude U_0^{loc} on S was obtained (Fig. 1). The autocorrelation orientational functions $P_1^{\|} = \langle a_{\|}(0) a_{\|}(t) \rangle / \langle a_{\|}^2 \rangle$ and $P_1^{\perp} = \langle a_{\perp}(0) a_{\perp}(t) \rangle / \langle a_{\perp}^2 \rangle$ were calculated. $a_{\|}$ and $a_{\perp}$ are the projections of chain bonds on the direction of the field and on the normal direction, respectively.

The presence of the field leads to anisotropy of the local orientational mobility of chain elements. The anisotropy is due to two mechanisms of orientational motions of chain elements in the field: (i) motions within the potential well caused by the field (characteristic times $\tau^{\perp}$ of these motions decrease with the field increase; (ii) transitions over the potential barrier U_0 caused by the field. In this case the times $\tau^{\|}$ increase with the field increase.

It was shown in [1] that the effect of an internal field on the local mobility of chain bonds is similar to that of the external field. It permits one to simulate the motion of mesogenic groups in LC polymers as the motion of a single chain in the viscous medium in the presence of the effective orienting potential inducing the necessary value of S.

Two models of a polymer chain in a quadrupole field (1) have been studied by the method of Brownian dynamics. The first model is the freely joint chain with rigid bonds [2]. The field acts on

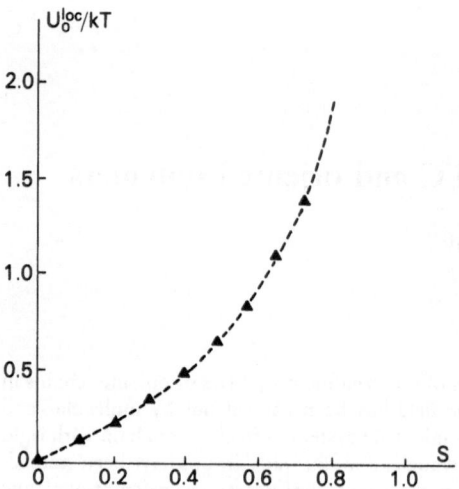

Fig. 1. Dependence of the local molecular field magnitude on S for a system of chains. The number of chain bonds $N = 5$, the full number of chains in the cell $n = 18$, the density $\rho = 0.8\,\rho_{max}$, the length of the bond $l = 0.69\sigma$, the temperature $T = 1.2\varepsilon_0/k$. ε_0 and σ are the parameters of the Lennard–Jones potential

chain bonds modeling the situation in polymer LC, consisting of chains with mesogenic groups in the main chain. The second model is the chain with fixed bond angles, free internal rotation and side groups similar to chain bonds. The field acts only on the side groups rigidly connected to the backbone [3, 4]. This is the simplest model for comb-like polymers in nematic state or solution of such chains in mesogenic solvent.

For both models connections between elements practically do not influence the equilibrium value of S for these elements. In the first case chains are oriented along the field direction. In the second case the chain conformation becomes oblate in the plane normal to the field direction with the side groups ordering. The dependences of orientational relaxation times $\tau^{\parallel}(S)$ and $\tau^{\perp}(S)$ for chain elements (bonds or side groups) are similar to those for such elements not connected with the chain.

It is shown that the dependence of the transition time τ_t on the barrier height U_0 is weaker than that predicted by the Kramers model [5]. This is due to the two-dimensional nature of the potential barrier. By using the stationary flow method, we obtained an expression for τ_t,

$$\tau_t = \tau_0/(U_0/kT)^{3/2}\exp(U_0/kT)$$

[6], which is in excellent agreement with the results of computer simulation and with the results of numerical solution of the diffusion equation, obtained by Martin et al. [7].

The normal-modes relaxation spectrum for a freely joint chain splits into two branches: (i) for motions which change the projections of the elements on the field direction, and (ii) for motions which change the projections on the normal direction. The branches have different dependences of the relaxation times on the motion scale and the field intensity U_0.

These dependences are in good agreement with the results of approximate solutions for the visco-elastic model in the quadrupole field, developed in [8] (Fig. 2). In this model the rigidity of chain bonds was kept "in average" by introducing elastic constants depending on S. For the second model the main-chain relaxation spectrum gets narrower with the side groups ordering (Fig. 3). This effect is similar to the effect of internal friction for visco-elastic chain models.

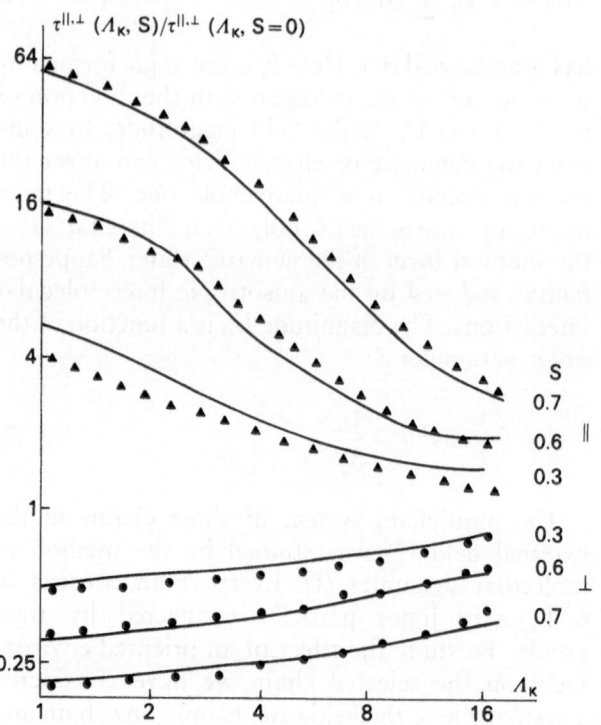

Fig. 2. Dependence of normal-modes relaxation times $\tau_k^{\parallel}$ and $\tau_k^{\perp}$ on the scale of motion $\Lambda_k = N/k$. $N = 16$ is the number of bonds, k is the mode number. Solid lines represent the analytical results of [8]

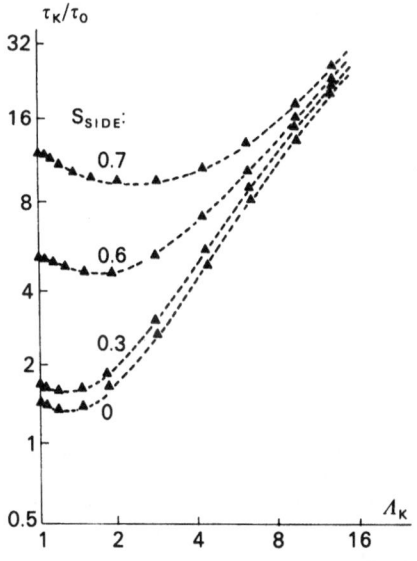

Fig. 3. Dependence of main-chain normal-modes relaxation times on the scale of motion for the model with side groups for different values of the side group order parameter S_{side}

References

1. Neelov I, Darinskii A, Lukyanov M, Gotlib Yu (1981) In: Molecular mobility in polymer systems. Abstracts of 12th Europhysics Conference on Macromolecular Physics, Leipzig, p 347
2. Darinskii A, Gotlib Yu, Lyulin A, Neelov I (1991) Vysocomol soed (USSR) A33:1211–1220
3. Darinskii A, Gotlib Yu, Lyulin A, Klushin L, Neelov I (1990) Vysocomol soed (USSR) A32:2289–2295
4. Darinskii A, Lyulin A, Neelov I (1992) Vysocomol soed (USSR) A34:73–83
5. Kramers H (1940) Physica 7:284
6. Lyulin A, Darinskii A, Gotlib Yu (1992) Physica A 182:607–616
7. Martin A, Meier G, Saupe A (1971) Symp Faraday Soc 5:119–133
8. Gotlib Yu (1989) Progr Colloid Polym Sci 80:245–253

Authors' address:

A. Darinskii
Institute of Macromolecular Compounds
199004, Bolshoi pr. 31
St. Petersburg, Russia

Local dynamics of freely rotating polymer chains in dense systems

B. Erman and I. Bahar

Polymer Research Center, Bogazici University, Istanbul, Turkey

Abstract: Local dynamics of a chain in the bulk state is strongly controlled by the immediate environment of the chain. The excessive friction exerted by the environment precludes large scale kinetic displacements of the chain atoms. In the presence of strong constraining effects of the environment, a transition from one configuration to the other is expected to take place by strongly correlated and relatively small range rearrangements of torsional motions. In the present study, local dynamics of the chain is analyzed as a succession of configurational rearrangements such that the change in the chain radius of gyration at each step is minimum. The degree of cooperativity among neighboring bonds along the chain resulting from the constraint of the environment is found to be highly localized. A 120° rotation of an internal bond is observed to be possible by cooperative adjustments of a few bonds in its immediate neighborhood along the chain. Changes in the mean-square anisotropic polarizability tensor when the ends of a relatively short chain ($n \leq 50$) are subject to a prescribed deformation, are also calculated. Calculations show that changes in the mean-square anisotropic polarizability depend strongly on chain length.

Key words: Local chain dynamics – restrictive environment – freely rotating chain – dense systems, anisotropic polarizability – configurational transitions, chains with fixed ends

Introduction

The transition of a polymer chain from one configuration to another results primarily from torsional fluctuations of bond rotation angles. The extent and nature of these fluctuations depend significantly on the state of the environment of the chain. The single chain surrounded by small solvent molecules is one extreme case where the torsional fluctuations are least constrained. At the local scale covering a few bonds, i.e. length scales of about 10–20 Å, the rate of spatial rearrangements of the chain, involving rotameric transitions of one or more backbone bonds, are affected both by bond torsional energy barriers and the friction exerted by the surrounding solvent, in addition to the intrinsic chain connectivity effect. The types of transitions complying with these constraints have been discussed previously [1, 2]. The local dynamics of chains resulting from such large torsional transitions has recently been studied by the dynamic rotational isomeric state model [3]. The bulk state constitutes the other extreme case in which the dynamics is strongly controlled by the environment and the torsional energy barriers are of secondary importance. The excessive friction exerted by the environment precludes large-scale displacements of the chain atoms. A transition from one configuration to another is, therefore, expected to take place by strongly correlated and relatively small rearrangements of torsional angles. In this case, the local dynamics of the chain may be analyzed as a succession of configurational rearrangements such that the change in the chain radius of gyration at each step is minimum. In the present communication, we discuss the degree of cooperativity among neighboring bonds resulting from the constraints of the environment and calculate the changes in the anisotropic polarizability tensor when the ends of a chain are subject to a prescribed deformation.

Theory

Formulation of the problem

We assume that the chain consists of n freely rotating bonds. The first atom is referred to as the zeroth atom. The notation follows that of Flory [4]. The change in the torsional angle of the ith bond is $\Delta\varphi_i$. The vector r_i denotes the position vector of the ith atom relative to a laboratory-fixed coordinate system OXYZ. The vector s_j from the mass center to the ith atom may be written in differential form as

$$\Delta s_i = \Delta r_i - (n+1)^{-1} \sum_{j=0}^{n} \Delta r_j . \tag{1}$$

The squared displacement of the ith atom is obtained from Eq. (1) as

$$(\Delta s_i)^2 = \left[(n+1)^2 \, \Delta r_i \cdot \Delta r_i + \sum_{j=3}^{n} \sum_{k=3}^{n} \Delta r_j \cdot \Delta r_k \right.$$
$$\left. - 2(n+1) \sum_{j=3}^{n} \Delta r_i \cdot \Delta r_j \right] \Big/ (n+1)^2 . \tag{2}$$

According to the basic postulate of the present study, the conformational rearrangements of the backbone atoms following an external perturbation are constrained to occur in a concerted fashion so as to preserve as much as possible their original position vectors s_i with respect to the center of gravity of the chain segment. Mathematically, this requirement is satisfied by minimizing the scalar function S which is defined as

$$S \equiv \sum_{i=3}^{n} (\Delta s_i)^2$$
$$= \sum_{i=3}^{n} \sum_{j=3}^{n} \left\{ \left[\delta_{ij} - \frac{(n+4)}{(n+1)^2} \right] \Delta r_i \cdot \Delta r_j \right\} . \tag{3}$$

Here δ_{ij} is the Kronecker delta. The second equality follows from the summation of Eq. (2) over the index i.

The motivation for minimizing the function S given in Eq. (3) rests on the idea of "compensating motions" of bonds, previously discussed by Skolnick and Helfand [5]. Accordingly, the bonds in the vicinity of a rotating bond rearrange themselves in a cooperative way such that excessive translations of points along the chain are avoided. The minimization of the function S inherently results from the cooperativity between bonds in ac-

comodating a given external perturbation. Results of calculations presented below indeed indicate that the rotation of an internal bond by an amount $\Delta\varphi$, induces counterbalancing rotations in at least one of the closest neighboring bonds on each side. The minimization of the S function may be regarded in this respect as a first-order approximation to the solution of the problem.

For simplicity, we assume that the chain is localized in space by fixing one end of it. This eliminates six variables representing the position of the zeroth atom and the absolute orientation of the chain in space. Rotations about the terminal bonds are not defined because of the absence of a preceding or succeeding bond with respect to which they would be defined. Thus, a given configuration of an n bond chain segment, may be described in terms of a set of $n-2$ torsional angles, from φ_2 to φ_{n-1}. The angles θ between adjacent bonds and the bond lengths l, are assumed to be the same for each bond and fixed throughout the complete dynamics. The vector r_i may be expressed in terms of the torsional angles φ_i as [4]

$$r_i = (E + T_1 + T_1 T_2 + \ldots + T_1 T_2 \ldots T_{i-1}) I \tag{4}$$

where E is the third-order identity matrix, I is the first-bond vector expressed in the frame OXYZ, and T_i is the conventional transformation matrix [4] expressing the bond $i+1$ in the frame of the ith bond. The displacement Δr_i of the ith atom, due to small angular changes $\Delta\varphi_j$, $2 \leq j \leq n-1$, is given by the following differential form

$$\Delta r_i = T_1 A_2 (E + T_3 + T_3 T_4$$
$$+ \ldots T_3 T_4 \ldots T_{i-1}) l \, \Delta\varphi_2$$
$$+ T_1 T_2 A_3 [E + T_4 + T_4 T_5 + \ldots$$
$$+ (T_4 T_5 \ldots T_{i-1})] l \Delta\varphi_3 \ldots$$
$$+ T_1 T_2 T_3 \ldots T_{i-2} A_{i-1} l \, \Delta\varphi_{i-1} \tag{5}$$

where, $A_j \equiv dT_j/d\varphi_j = AT_j$ with

$$A = \begin{bmatrix} 0 & 0 & 0 \\ 0 & 0 & -1 \\ 0 & 1 & 0 \end{bmatrix} . \tag{6}$$

Equation (5) may be expressed in compact form as

$$\Delta r_i = \sum_{j=2}^{i-1} a_{ij} \Delta\varphi_j \tag{7}$$

where for each i, $\mathbf{a}_{ij}$ is a 3×1 vector defined as

$$\mathbf{a}_{ij} = \mathbf{T}_1^{(j-1)} \mathbf{A} \left\{ [\mathbf{E} \quad 0] \mathbf{G}_j^{(i-1)} \begin{bmatrix} 1 \\ 1 \end{bmatrix} - 1 \right\} . \qquad (8)$$

Here, the notation $\mathbf{T}_1^{(j-1)}$ represents the product of transformation matrices, from $\mathbf{T}_1$ to $\mathbf{T}_{j-1}$, and $\mathbf{G}_j$ is the generator matrix given by

$$\mathbf{G}_j = \begin{bmatrix} \mathbf{T}_j & \mathbf{I} \\ 0 & 1 \end{bmatrix} . \qquad (9)$$

The partial derivative of S with respect to the angular displacement, $\Delta\varphi_m$, is obtained from Eq. (4) as

$$\frac{\partial S}{\partial \Delta\varphi_m} = 2 \sum_{i=3}^{n} \Delta\mathbf{r}_i \frac{\partial \Delta\mathbf{r}_i}{\partial \Delta\varphi_m}$$
$$- \frac{(n+4)}{(n+1)^2} \sum_{i=3}^{n} \sum_{j=3}^{n} \left(\Delta\mathbf{r}_i \cdot \frac{\partial \Delta\mathbf{r}_j}{\partial \Delta\varphi_m} \right.$$
$$\left. + \Delta\mathbf{r}_j \cdot \frac{\partial \Delta\mathbf{r}_i}{\partial \Delta\varphi_m} \right) , \qquad (10)$$

where $\partial \Delta\mathbf{r}_i / \partial \Delta\varphi_m = \mathbf{a}_{im} \mathbf{H}(m,i)$ with $\mathbf{H}(m,i)$ being equal to unity for $m < i$ and to zero for $m \geq i$. Equation (10) becomes

$$\frac{1}{2} \frac{\partial S}{\partial \Delta\varphi_m} = \sum_{i=3}^{n} \mathbf{a}_{im} \mathbf{H}(m,i) \cdot \sum_{j=2}^{i-1} \mathbf{a}_{ij} \Delta\varphi_j$$
$$- \frac{(n+4)}{(n+2)^2} \mathbf{v}_m \cdot \sum_{i=3}^{n} \sum_{j=2}^{i-1} \mathbf{a}_{ij} \Delta\varphi_j , \qquad (11)$$

where

$$\mathbf{v}_m \equiv \sum_{i=3}^{n} \mathbf{a}_{im} \mathbf{H}(m,i) . \qquad (12)$$

Equation (11) may be cast in a simpler form by substituting the following definition

$$u_{mj} = \sum_{i=3}^{n} \mathbf{a}_{im} \mathbf{H}(\mathbf{m,i}) \cdot \mathbf{a}_{ij} \mathbf{H}(j,i) \qquad (13)$$

and rearranging. This leads to

$$\frac{1}{2} \frac{\partial S}{\partial \Delta\varphi_m} = \sum_{j=2}^{n-1} \left[u_{mj} - \frac{(n+4)}{(n+1)^2} \mathbf{v}_m \cdot \mathbf{v}_j \right] \Delta\varphi_j$$
$$= \sum_{j=2}^{n-1} R_{mj} \Delta\varphi_j , \qquad (14)$$

where R_{mj} is defined from the second equality of Eq. (14) as

$$R_{mj} = u_{mj} - \frac{(n+4)}{(n+1)^2} \mathbf{v}_m \cdot \mathbf{v}_j . \qquad (15)$$

The solution of the problem proceeds by equating the expression given by Eq. (14) to zero with the proper constraints imposed on the motion. For illustrative reasons, we consider the case where the zeroth atom is fixed at the origin of the coordinate system, the nth atom at the other end of the chain is displaced by an amount $(\Delta x^0, \Delta y^0, \Delta z^0)$ and the torsional angle $\Delta\varphi_{\mathrm{mid}}$ of the middle bond is changed by an amount $\Delta\varphi^0$. We search for the set $\{\Delta\varphi_m\}$ of torsional displacements of the internal bonds m, $2 \leq m \leq n-1$, excluding $\Delta\varphi_{\mathrm{mid}}$, that will satisfy the requirement of minimum displacement of position vectors relative to the mass center, subject to the condition that the end of the chain and the central bond torsional angles are displaced by the prescribed amounts. The problem therefore reduces to the minimization of S in the presence of four Lagrange multipliers as

$$\frac{1}{2} \frac{\partial}{\partial \Delta\varphi_m} [S - \lambda_x(\Delta x_n - \Delta x^0) - \lambda_y(\Delta y_n - \Delta y^0)$$
$$- \lambda_z(\Delta z_n - \Delta z^0)$$
$$- \lambda_\varphi(\Delta\varphi_{\mathrm{mid}} - \Delta\varphi^0) = 0 . \qquad (16)$$

Here, λ_x, λ_y, λ_z and λ_φ are the four Lagrange multipliers, Δx^0, Δy^0 and Δz^0 are the components of the imposed displacement of the right end of the chain, and Δx_n, Δy_n and Δz_n are the three components of the displacement of the nth atom and $\Delta\varphi_{\mathrm{mid}}$ is the torsional rotation of the middle bond. Equation (16) represents a set of $n+1$ equations and $n+1$ unknowns. The unknowns are the elements of the set $\{\Delta\varphi_m\}$, $n-3$ in number – the known $\Delta\varphi_{\mathrm{mid}}$ is excluded from the set – and the four Lagrange multipliers. Ensemble averages are obtained by a Monte Carlo scheme over different initial configurations.

Calculations

The degree of cooperativity among neighboring bonds of a chain may be studied by rotating the middle bond by a prescribed angle $\Delta\varphi^0$ and calculating the resulting rotations over the other bonds. The solution is obtained from Eqs. (14)–(17) by setting Δx^0, Δy^0 and Δz^0 to zero and solving the set of linear equations for the fourth Lagrange multiplier λ_ϕ and the rotations $\Delta\varphi_i$. In Fig. 1 the results of calculations are shown for a given initial configuration of a chain of 25 bonds. The initial conformation is chosen by assigning a random rotational

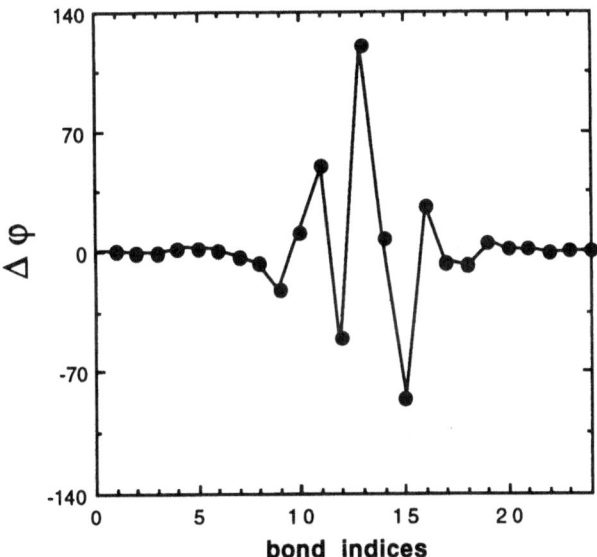

Fig. 1. Change in dihedral angles following a change of 120° in the torsional angle of a chain of 25 bonds, as a function of bond index

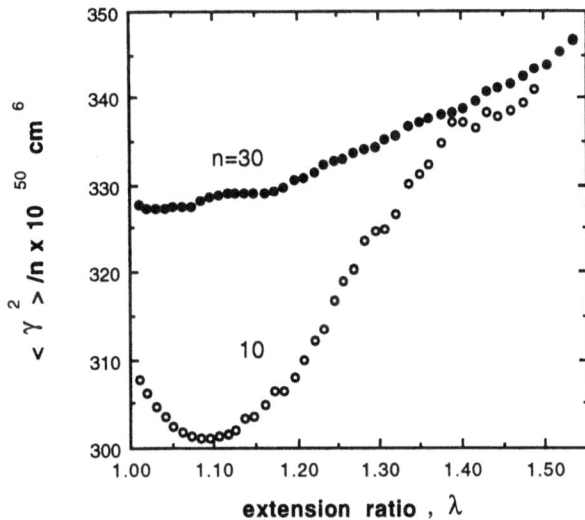

Fig. 2. Change in mean-square anisotropy of the polarizability tensor as a function of extension ratio for two short chains of 10 and 30 bonds

isomeric state to each bond, as *trans*, gauche$^+$ or gauche$^-$. The 13th bond is rotated by an amount of 120° from its original value, with increments of 2°. The immediate neighbors on both sides respond by a counterrotation of about 70°. The second neighbors on both sides respond by about 50° rotations in the same direction as the 13th bond. The responses of the third and more distant neighbors are very small. Although Fig. 1 is representative of a given configuration only, numerous calculations with different initial configurations show that a total of about seven bonds respond cooperatively and bonds farther away from this group are insignificantly affected. We may conclude, therefore, that the rotation about a given bond is highly localized by the coupled spatial rearrangements of its near neighbors along the chain. This local cooperativity, resulting from the linear connectivity of the chain, is confined to about seven bonds.

In the second set of calculations, the changes in the anisotropy of the chain polarizability tensor under deformation are obtained. The mean-square anisotropy $\langle \gamma^2 \rangle$ is defined as

$$\langle \gamma^2 \rangle = (3/2)\text{trace}\langle \hat{\alpha}\hat{\alpha} \rangle , \qquad (17)$$

where the right-hand side follows from the definition given by Flory [4]. In Fig. 2, values of $\langle \gamma^2 \rangle$ are shown for chains of 10 and 30 bonds. Polarizability values for polyethylene are adopted in the calculations. Points in Fig. 2 represent averages over 500

chains of different initial configurations obtained by Monte Carlo technique. One end of each chain is held fixed and the other is stretched by an extension ratio of λ. The value of γ^2 is calculated for each chain and averaged over all chains generated. It is interesting to note that the anisotropy decreases for the chain of 10 bonds at lower degrees of extension and gradually converges to that of the 30 bond chain at higher stretch ratios. This indicates that the coefficient of strain anisotropy depends sensitively on chain length and chain conformational characteristics.

References

1. Helfand E (1971) J Chem Phys 54:4651
2. Valeur B, Jarry JP, Geny F, Monnerie L (1975) J Polym Sci Polym Phys Ed 13:667
3. Bahar I, Erman B (1987) Macromolecules 20:1368
4. Flory PJ (1969) Statistical Mechanics of Chain Molecules. Wiley-Interscience, New York
5. Skolnick J, Helfand E (1980) J Chem Phys 72:5489

Received January 22, 1992;
accepted June 5, 1992

Authors' address:

Burak Erman
Polymer Research Center and School of Engineering
Bogazici University
Bebek 80815
Istanbul, Turkey

Progress in Colloid & Polymer Science   Progr Colloid Polym Sci 91:20–23 (1993)

Relaxation behaviour in bulk PIMA and PIMA–PMMA copolymer near T_g

F. Alvarez[1]), J. Colmenero[1]), C. H. Wang[2]), and G. Fytas[3])

[1]) Departamento de Física de Materiales, Universidad del País Vasco, San Sebastián, Spain
[2]) Department of Chemistry, University of Nebraska, Lincoln, Nebraska, USA
[3]) Forth-Iesl, Heraklion, Crete, Greece

Abstract: The relaxation processes in the glass-transition temperature range of bulk PIMA and PIMA–PMMA copolymer have been studied by means of Photon Correlation Spectroscopy (PCS) and dielectric spectroscopy (DR). The experimental results were treated in a consistent way employing an inverse Laplace transform analysis of the density $C_\rho(t)$ and dipole moment $\varphi_\mu(t)$ autocorrelation functions. The latter was obtained from the $\varepsilon''(\omega)$ data utilizing a recently developed algorithm. For bulk PIMA, $C_\rho(t)$ clearly shows two distinct relaxation processes, whereas $\varphi_\mu(t)$ is dominated by the slower (α-) relaxation mode. The fast process in $C_\rho(t)$ is probably related to the rotation of the rigid isobornyl group about the C–O bond.

The copolymer displays a β-relaxation process and a slower α-relaxation dielectric process in agreement with the PCS results. The molecular origin of these processes is outlined.

Key words: PIMA – PMMA – PCS – dielectric spectroscopy – relaxation

Introduction

Due to its transparency in the visible region and desirable mechanical properties, amorphous atactic polymethyl methacrylate (PMMA) is currently used for fiber optics application. The glass transition temperature (T_g) of atactic PMMA is about 100 °C. As a result, PMMAs fiber optics applications are limited to about 80 °C, above which PMMA fiber becomes soft and loses its mechanical integrity. However, by replacing the methyl group in the ester moiety with a bulky bicyclic isobornyl group [1], one obtains polyisobornyl methacrylate (PIMA), which has a glass transition temperature as high as 150 °C in the asymptotic high molecular weight limit. The high T_g of PIMA is due in part to the difficulty of undergoing free rotation of the isobornyl side group, which results in reduced chain flexibility.

In addition to high T_g, PIMA is heat-resistive, and also exhibits superior optical transparency. However, the results of our extensive investigations of mechanical and optical properties of PIMA and PIMA/PMMA copolymers show that pure PIMA has a considerably smaller shear modulus and is more brittle than PMMA [2]. But, by blending or copolymerizing PIMA with PMMA, the mechanical property is considerably improved, yet without significantly effecting the optical property. This is due to the fact that the refractive index of PIMA is practically the same as that of PMMA. Consequently, the intensity loss due to light scattering for fibers made of a PIMA/PMMA blend or copolymer remains unchanged as compared to pure PMMA. The light scattering loss arising from composition inhomogeneity (or concentration fluctuations) is insignificant in the PIMA/PMMA blend and copolymer.

However, despite the potential for fiber optics application at higher temperature for PIMA/PMMA copolymers, there appears to be no report available in the scientific literature on the physical characterization data of PIMA or PIMA/PMMA copolymer in the bulk state. To elucidate the motion involved in PIMA and PIMA/PMMA copolymer, we have used photon correlation spectroscopy (PCS) and dielectric relaxation (DR) to investigate the dynamics of segmental motion of the main chain and the reorientation of the ester side group that bear the bulky isobornyl moiety. The result is compared with that obtained for bulk PMMA [3] previously published.

Experimental

The PCS measurements were performed with an argon ion laser (Spectra Physics 2020) operating at a wavelength of $\lambda = 488$ nm with a power of 150 mW. Intensity autocorrelation functions, $G(t)$, were measured in the time range 10^{-6}–10^2 s on a ALV-5000 correlator. For amorphous polymers which display homodyne scattering above T_g the density time correlations function $g(t)$ is related to $G(t)$ by $G(t) = A(1 + f|\alpha g(t)|^2)$. The temperature range measured was from 418 to 438 °C for bulk PIMA and 403 to 453 K for the copolymer. As an example Fig. 1 shows the correlogram for the copolymer at 423 K.

Dielectric spectroscopy measurements of the susceptibility, $\chi^* = \chi' - i\chi''$ were performed by means of a lock-in amplifier EG&G PAR 5208 scanning the frequency in the range 5–10^5 Hz at isothermal conditions between 420 and 470 K for bulk PIMA and between 320 and 440 K for the copolymer. Scans on temperature were also performed down to 100 K to search for low-temperature relaxations. (Fig. 2).

The glass transition temperatures T_g of the samples as measured by Differential Scanning Calorimetry at a heating rate of 10 K/min were 396 K for bulk PIMA and 380 K for the copolymer.

Data analysis

Traditional data analysis is based upon the supposition of assuming Havriliak–Negami functions for the dielectric loss peaks. This method of analysis poses the main problem of the need of too many parameters whenever more than a single process appears in the frequency window as it happens to be in this work. Moreover, comparison between dielectric and PCS results is not straightforward by this way as long as PCS is a time-domain spectroscopy, whereas DR is frequency-domain [4].

We have developed a new analysis which makes use of an iterative algorithm [5], which yields a distribution of relaxation times, $\rho(\ln \tau)$, [Eq. (1)] which is highly performant from a mathematical point of view in the sense that it is able to reproduce accurately the data, but it is not easy to understand from the physical point of view.

$$\frac{\varepsilon''(\omega)}{\varepsilon_s - \varepsilon_\infty} = \int_{-\infty}^{\infty} \rho(\ln \tau) \frac{\omega\tau}{1 + \omega^2\tau^2} d\ln\tau . \quad (1)$$

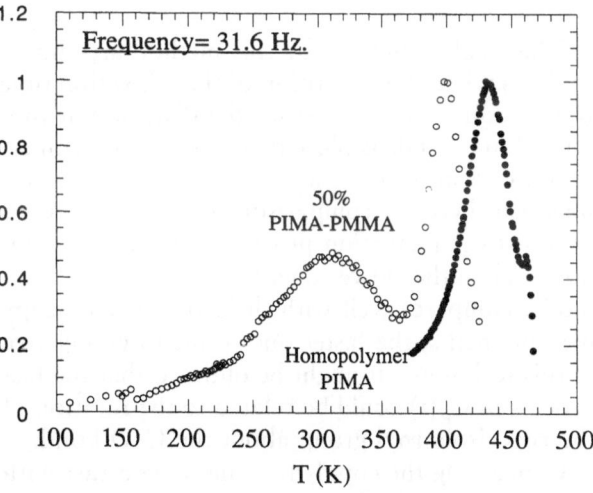

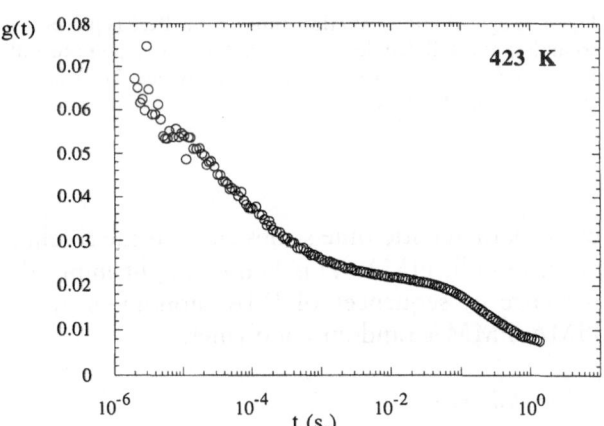

Fig. 1. Density autocorrelation function of the 50% PIMA–PMMA copolymer at 423 K

Fig. 2. Normalized imaginary part of the dielectric susceptibility at 31.6 Hz as a function of temperature. Solid symbols correspond to bulk PIMA and the open ones are for the copolymer

On a next step, this so obtained distribution is integrated in order to obtain the time relaxation function, $\varphi(t)$ [Eq. (2)].

$$\varphi(t) = \int_{-\infty}^{\infty} \rho(\ln \tau) \exp\left(-\frac{t}{\tau}\right) d\ln\tau \ . \qquad (2)$$

Afterwards, this time relaxation function is analysed with CONTIN [6], which gives a set of solutions from which we can choose at least one which can compare favourably with PCS CONTIN output. The consistency of this method is tested by calculating back the dielectric loss peak out of the chosen CONTIN solution and comparing it with the original data. To improve accuracy this process can be iterated: the CONTIN distribution can be inserted as an initial trial for the iterative algorithm, and then repeat all the procedure. Following this procedure comparison between dielectric and PCS data can be directly made on the grounds of distribution functions of relaxation times. As an example, Fig. 3 shows the normalized distribution functions obtained for the 50% sample at 414 K by dielectric spectroscopy and at 423 K by PCS, which correspond to the clearest and nearest pair of temperatures we can choose due to the limitations of the dynamical window in the case of PCS and to the appearance of the d.c. conductivity in the DRS measurements.

Results and discussion

The results obtained by this method are shown in Fig. 4 where the logarithm of the relaxation time (the maxima extracted from the distribution functions) in seconds is plotted versus inverse temperature. PCS measurements of bulk PMMA are also included there for comparison. Bulk PIMA shows two distinct relaxation processes in the PCS window. While the slower one, likely the α-relaxation mode, compares well with dielectric spectroscopy measurements, the faster one seems to be inactive in this technique. It might be outlined that the fast process in $C_\rho(t)$ could be related to the rotation of the rigid isobornyl group about the C–O bond.

Concerning the copolymer, and in contrast with PIMA, a dielectrically active β-relaxation process is detected which seems that can be assigned to the bulk PMMA β-relaxation. Moreover, $C_\rho(t)$ shows strong evidence of a second α-relaxation process

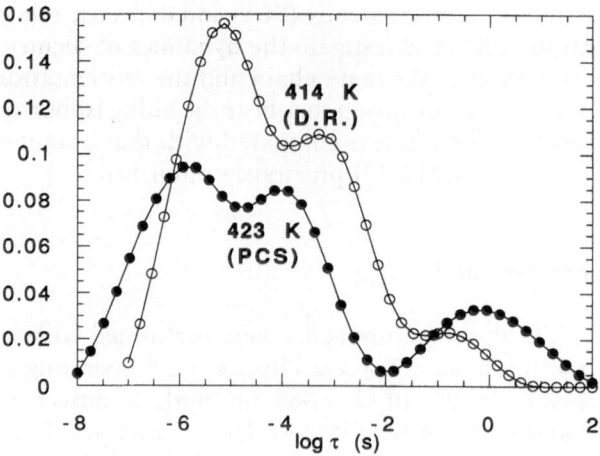

Fig. 3. Time relaxation distributions obtained as indicated in the text. Solid symbols are for PCS data and the open ones are from dielectric measurements

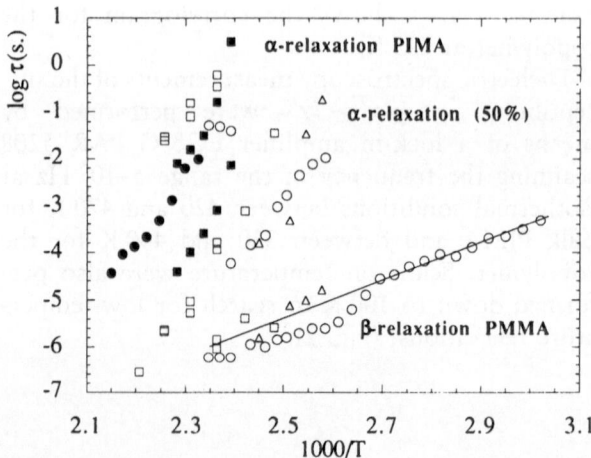

Fig. 4. Logarithmic relaxation times from PCS (squares and triangles) and DR (circles). Solid symbols correspond to bulk PIMA and the open ones are for the copolymer. The triangles stand for bulk PMMA

with characteristic time scales close to the α-relaxation in bulk PIMA. This finding might imply the presence of sequences of IMA monomers in the PIMA–PMMA random copolymer.

Acknowledgements

This work was supported by the CICyT and by Gipuzkoako Foru Aldundia. One of us (F. Alvarez) also acknowledges the grant of the Basque Government.

References

1.  a schematic view of the radical group corresponding to IMA

2. Xia JL, Wang CH, unpublished
3. Fytas G, Wang CH, Fischer EW (1988) Macromolecules 21:2253
4. Alvarez F, Alegría A, Colmenero J (1991) Phys Rev B44:7306
5. Imanishi Y, Adachi K, Kotaka T (1988)) J Chem Phys 89:7593
6. Provencher SW (1982) Comput Phys Commun 27:213

Received January 16, 1992;
accepted June 9, 1992

Authors' address:

J. Colmenero
Departamento de Física de Materiales, UPV/EHU
Apdo. 1072
20080 San Sebastián, Spain

Progress in Colloid & Polymer Science   Progr Colloid Polym Sci 91:24–27 (1993)

Dynamics of the α-relaxation in glass-forming polymeric systems. Study by neutron scattering and relaxation techniques

A. Arbe, A. Alegría, F. Alvarez, J. Colmenero, and B. Frick[1])

Departamento de Física de Materiales, Universidad del País Vasco, San Sebastián, Spain
[1]) Institute Laue-Langevin (ILL), Grenoble, France

Abstract: The dynamics of the α-relaxation in three glass-forming polymeric systems has been studied by means of quasielastic neutron scattering as well as by relaxation techniques. By using these techniques, we have covered a wide time-scale ranging from mesoscopic to macroscopic times (10^{-10}–10^1 s). All relaxation data have been interpreted in terms of a Havriliak–Negami relaxation function, Φ_{HN}. Neutron scattering data were described in terms of a scattering law $S(Q, \omega)$ which was also built starting from Φ_{HN}. The results obtained indicate that the dynamics of the α-relaxation in a wide timescale can be well described by means of the same spectral shape, i.e. the same Havriliak–Negami parameter values. Moreover, the Havriliak–Negami characteristic times deduced from the fitting of the experimental data can also be described using only one Vogel–Fulcher functional form. Then, this implies a self-consistent description of the dynamics of the α-relaxation obtained by very different probes.

Key words: Dynamics of the α-relaxation – glass transition – glass-forming polymeric systems – quasielastic neutron scattering – relaxation techniques

Introduction

During recent years, a great deal of effort has been made in the study of the dynamics of the α-relaxation in very different glass-forming systems (see as general references [1, 2]). From an experimental point of view, the main magnitudes characterizing the dynamics of the α-relaxation are the different relaxation times associated with the different probes as well as the shape of the relaxation function. Nowadays, it is well established experimentally that the relaxation times follow a non-Arrhenius temperature behavior which can be parametrized by means of different functional forms. Moreover, the shape of the relaxation function shows a clear non-Debye behavior. Almost all of these results have been obtained by means of standard relaxation techniques, like dielectric and mechanical spectroscopies or, in some cases, by

means of photon correlation spectroscopy, a light scattering technique in the time domain. All of these techniques cover a timescale ranging from about 10^{-5} to 10^2 s, i.e., a macroscopic range. Then a first question open to discussion is whether the results obtained by the above-mentioned techniques can be extended or not to a more microscopical timescale, like the mesoscopic range (10^{-11}–10^{-7} s). This timescale can be covered by incoherent quasielastic neutron scattering (QENS). In addition, this technique allows us to learn about how the magnitudes characterizing the dynamics of the α-relaxation depend on Q (where Q is the modulus of the change of the wavevector in a scattering experiment) in a Q-range roughly $0.2 < Q < 5 \, \text{Å}^{-1}$.

In this work we summarize the results obtained by means of QENS on the dynamics of the α-relaxation of three different polymeric systems.

These results are compared to the corresponding ones obtained by relaxation techniques in the same polymers.

Experimental

The samples investigated were poly(vinyl methyl ether) (PVME), poly(vinyl chloride) (PVC) and poly(bisphenol A, 2-hydroxypropylether) (PH). These glass-forming polymeric systems have the advantage that they do not crystallize, allowing us to explore a wide temperature range above the glass-transition temperature, T_g, without the problems of crystallization. However, for PVC we are limited both in temperature and measuring time due to the known tendency of PVC to degrade chemically above T_g. We have checked by dielectric spectroscopy that the thermal treatment followed during the neutron scattering measurements does not noticeably affect the shape and temperature behavior of the relaxation function. These results will be published elsewhere. The glass-transition temperatures were determined by differential scanning calorimetry (DSC), resulting to be $T_g = 250$ K (PVME), $T_g = 350$ K (PVC) and $T_g = 358$ K (PH).

QENS measurements were carried out by means of the neutron backscattering spectrometers IN10 and IN13 at the Institute Laue Langevin (ILL) in Grenoble. The incident wavelengths used by us were $\lambda = 6.28$ Å (IN10) and $\lambda = 2.23$ Å (IN13), giving an energy resolution of $\delta E \sim 1$ μeV and $\delta E \sim 8$ μeV, respectively. The Q-range covered was roughly 0.2–2 Å^{-1} on IN10 and 0.2–5.4 Å^{-1} on IN13. The samples (thickness: 0.15 mm) were filled in a cylindrical Al container yielding a transmission of about 85%. Typical measuring time at each temperature was 24 h for IN10 and 36 h for IN13. Initial data treatment was carried out in the normal way, correcting for effects of detector efficiencies, scattering from sample container, and instrumental background. The experimental curve for the incoherent scattering function $S(Q, \omega)$ was finally obtained at each temperature as a function of the frequency change on scattering, ω, and the modulus of the change of wavevector, Q.

Results

Due to the fact that we used protonated samples we mainly observed the incoherent scattering arising from the self-correlation function which involves the motion of protons. In the α-relaxation range, the dynamics of the polymers investigated was detected by neutron scattering as a quasielastic broadening by IN10 and by IN13 at temperatures higher than $T_g + 50$ K. Figure 1 shows a typical spectra obtained by IN10 corresponding to one of the polymers investigated. As can be seen in the figure, the spectral shape is clearly not well described by one Lorentzian function, i.e., by the Fourier transformation of a simple exponential relaxation function in time. A stretched exponential or the Kohlrausch–Williams–Watts (KWW) function could be an appropriate representation of the relaxation in time. However, as an analytical Fourier transformation of the KWW function does not exist, we have followed the procedure we have developed for PVME in ref. [3]. We express the scattering function $S(Q, \omega)$ in terms of the imaginary part of the Havriliak–Negami (HN) relaxation function $\Phi_{HN}^*(\omega)$,

$$S_{HN}(Q, \omega) \propto -\frac{1}{\omega} \text{Im} \left[\Phi_{HN}^*(\omega) \right]$$

$$= -\frac{1}{\omega} \text{Im}(\{1 + [i\omega\tau_{HN}(Q, T)]^\alpha\}^{-\gamma}), \quad (1)$$

and fit the experimental data directly in frequency.

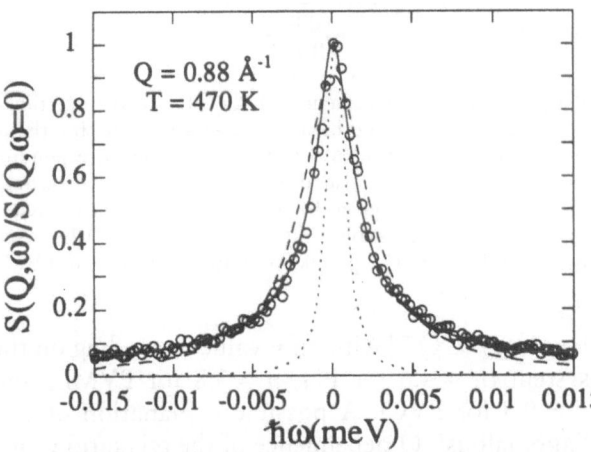

Fig. 1. Normalized $S(Q, \omega)$ of PH measured by IN10. The solid line is the fitting curve obtained through the $S_{HN}(Q, \omega)$, with $\alpha = 0.76$ and $\gamma = 0.42$, and the dashed line corresponds to a single Lorentzian fit. The measured resolution is also shown for comparison (dotted line)

Here, α and γ are two parameters in the range $(0 < \alpha, \gamma < 1)$ characterizing the spectral shape and τ_{HN} is a characteristic time of the relaxation process, depending on both, Q and temperature. We have shown in a previous paper [4] that the HN relaxation function can be considered as a good analytical description of the numerical Fourier transformation of the KWW function. In that paper we obtained, by numerical calculation, some empirical relationship between the HN parameters (α, γ, and τ_{HN}) and the corresponding KWW ones (β and τ_{WW}).

Constant spectral shape parameters (α, γ) were found to describe well the neutron scattering data, i.e., independent of temperature and the Q value, within experimental limitations. This finding agrees with the results from relaxation techniques, which gave as well a temperature-independent spectral shape for the polymers investigated. Moreover, for each polymer, the spectral shape parameters found were similar to the values measured by relaxation techniques [3, 5, 6]. Thus, the only Q- and T-dependent parameter seems to be the relaxation time $\tau_{HN}(Q, T)$. The values of $\tau_{HN}(Q, T)$ were determined from the fitting of the experimental data of IN10 and IN13 by means of Eq. (1), allowing in addition for a flat background (FBG). Possible explanations for the latter are widely discussed in ref. [3]. An example of the fitting procedure is shown in Fig. 1. The values of the α and γ parameters used in the fitting are shown in Fig. 2. The β values corresponding to the KWW function, which is related to the HN function, were calculated following the procedure reported in the above-mentioned paper [4]; they are also shown in Fig. 2.

Discussion

The Q-behavior of $\tau_{HN}(Q, T)$ found is almost independent of the temperature for the three polymers investigated, at least in the temperature range measured and inside the experimental errors involved. This implies that the Q-dependence and the temperature dependence of τ_{HN} may be factorized as $\tau_{HN}(Q, T) = a(T)\check{\tau}(Q)$, allowing to construct a master curve $\check{\tau}(Q)$, if the curves $\tau_{HN}(Q, T)$ vs. $\log Q$ for different temperatures are shifted by a factor $a(T)$ on the logarithmic τ-scale. Within experimental errors, we may describe the Q-dependence of the relaxation times, $\check{\tau}(Q)$, by a power

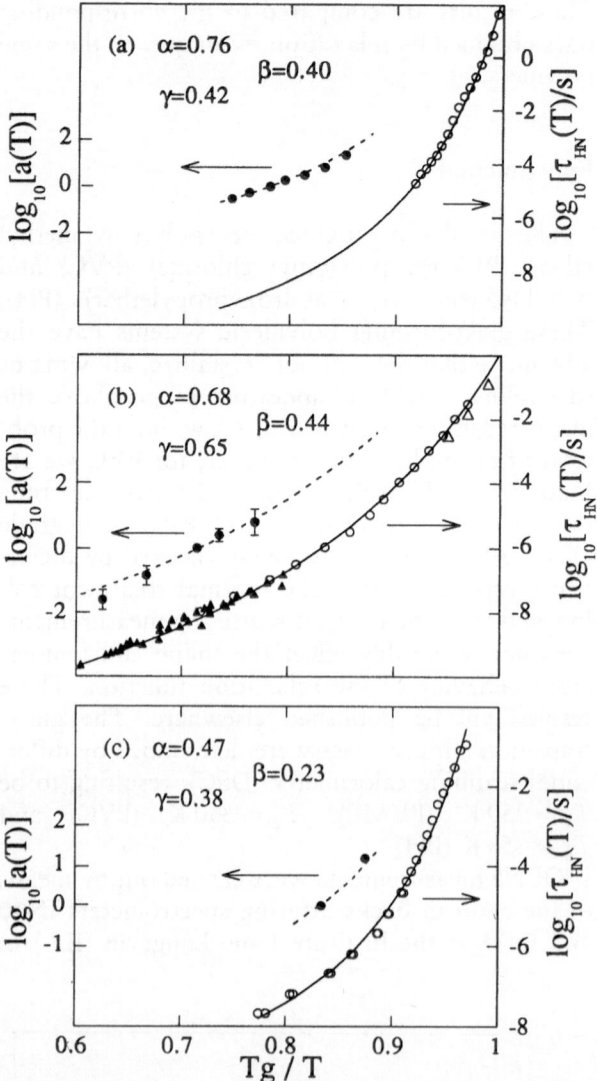

Fig. 2. Temperature behavior of the HN relaxation times, ($\bigcirc$) dielectric, ($\triangle$) mechanical, and ($\blacktriangle$) NMR, for the α-process of (a) PH, (b) PVME and (c) PVC. The corresponding shape parameters are also displayed. Dashed lines through $a(T)$ points ($\bullet$) are given by the same VF laws fitting relaxation data (solid lines) but shifted in the timescale. Characteristic error bars in $a(T)$ points are shown in the case of PVME

law $\check{\tau}(Q) \propto Q^{-n}$ with an n value depending on the system ($n = 4.6$ for PH, $n = 3.8$ for PVME, and $n = 9.5$ for PVC). A possible explanation of this "anomalous" Q-dependence of the relaxation time is discussed elsewhere [7].

On the other hand, the temperature dependence of $\tau_{HN}(Q, T)$, which is given by $a(T)$, has been plotted in Fig. 2, for the three polymers, in comparison to the temperature behavior of the HN relaxation times obtained from relaxation techniques. It is clear that the values of $\tau_{HN}(T)$ obtained by relaxation techniques have a temperature behavior very different from the corresponding one to an activated relaxation. As can be seen in Fig. 2, the Vogel–Fulcher (VF) law

$$\tau(T) \propto \exp[A/(T - T_0)]$$

fits very well the temperature dependence of the characteristic time τ_{HN} of the macroscopic α-relaxation. Moreover, as Fig. 2 displays, the same laws also perfectly fit the $a(T)$, i.e., the temperature dependence of the HN relaxation times, obtained from neutron scattering measurements.

Conclusions

Therefore, the main conclusion of this paper is that the dynamics above the glass-transition in a wide timescale (10^{-10}–10^1 s) can be well described by using the same "spectral shape" i.e., the same Havriliak–Negami parameters or the same equivalent β-value of the stretched exponential function. Moreover, the characteristic timescale τ_{HN} deduced from these fits can also be described by using only one Vogel–Fulcher functional form. Then, this implies a self-consistent description of the dynamics obtained by very different probes.

Acknowledgements

This work has been supported by the CICyT (project: MEC MAT89-0816) and by Gipuzkoako Foru Aldundia. One of us (A. Arbe) also acknowledges a grant of the Basque Government.

References

1. Richter D, Dianoux AJ, Petry W, Teixeira J (eds) (1989) Dynamics of Disordered Materials. Springer, Berlin
2. Ngai KL, Wright GB (eds) (1991) Proceedings of the International Discussion Meeting on Relaxations in Complex Systems held in Heraklion, Crete, Greece, June 1990, on Relaxations in Complex Systems, J Non-Crystalline Solids 131–133
3. Colmenero J et al. (1991) Phys Rev B 44:7321
4. Alvarez F, Alegría A, Colmenero J (1991) Phys Rev B 44:7306
5. Del Val JJ et al. (1989) Makromol Chem 190:7321
6. Alegría A et al., Phys Rev B, submitted
7. Colmenero J et al. (1992) Phys Rev Lett 69:478

Received January 16, 1991;
accepted June 9, 1992

Authors' address:

J. Colmenero
Departamento de Física de Materiales, UPV/EHU
Apdo. 1072
20080 San Sebastián, Spain

Progress in Colloid & Polymer Science

Progr Colloid Polym Sci 91:28–30 (1993)

Incoherent quasielastic neutron scattering study of a glass-forming liquid. A mode coupling interpretation

G. Floudas[1]), J. S. Higgins[1]), and A. Burgess[2])

[1]) Imperial College, Dept. of Chem. Eng., London, UK
[2]) ICI, Runcorn, Liverpool, UK

Abstract: We report on incoherent neutron scattering results in the glass-forming liquid di-2-ethylhexyl phthalate (DOP) ($T_g = 184$ K). We are particularly interested in the temperature dependence of (i) the elastic scattered intensity, (ii) the mean-square displacement and (iii) the shear viscosity. Our neutron scattering results are internally consistent and compatible with the existence of a dynamic instability at temperatures above T_g, as predicted by the recent mode coupling theory. This is in agreement with recent independent measurements on the same liquid from dielectric relaxation and depolarized Rayleigh scattering. However, an alternative approach [9] based on free-volume arguments shows no discontinuity over a broad T-range.

Key words: Glass transition – incoherent neutron scattering – di-2-ethylhexyl phthalate

Introduction

Recent mode coupling theory (MCT) predicts a dynamic instability at a critical temperature T_c, located above the calorimetric glass transition temperature T_g [1]. This critical behaviour has been observed in some "fragile" organic glass-forming liquids [2–4], polymeric systems [5] and a molten salt system [6], mainly with the use of coherent and incoherent neutron scattering. In a recent study [7] on the dynamics of the glass-forming liquid di-2-ethylhexyl phthalate (DOP) we found good agreement between depolarized Rayleigh scattering (DRS), dielectric relaxation (DR) and neutron scattering relaxation times. However, the Stokes–Einstein–Debye (SED) equation ($\tau \sim \eta/T$, where τ is the orientation relaxation time and η is the shear viscosity) was found to predict slower τ values at temperatures close to T_g. This resulted in the breakdown of the SED equation which was demonstrated in a plot of τ/η versus T_g/T, at $T_g/T \sim 0.7$–0.8. This peculiar behaviour of DOP associated with the orientational dynamics at temperatures well above T_g, that is, in a temperature range of interest to MCT, deserves more attention. Keeping in mind that the theory was made to describe simple liquids like hard spheres or Lennard–Jones spheres and given that DOP constitutes a simple liquid, it would be interesting to see if some of the MCT predictions agree with our neutron scattering data.

Experimental section

The neutron scattering experiments were carried out on the IRIS backscattering spectrometer at the ISIS pulsed neutron source at RAL. The energy resolution of the spectrometer was 15 μeV and the energy range from -0.4 to 0.4 meV. The Q-range covered was from 0.33 to 1.85 Å^{-1}. Details on the experimental conditions can be found in ref. [7].

Theoretical background

Among the theoretical predictions of MCT of fundamental importance is the existence of two distinct relaxation processes (α and β) in the density correlation function, on the liquid side of the transition. Near T_c, the two processes display power law behaviour with critical exponents. More specifically, MCT predicts that the long-time limit of the β-process or the anomalous Debye–Waller factor $f_Q(T)$ shows a cusp behaviour:

$$f_Q(T) = \begin{cases} f_Q^c + h_Q C \sqrt{\dfrac{T_c - T}{T_c}} + O(T), & T \leq T_c, \\ f_Q^c + O(T), & T \geq T_c, \end{cases}$$

(1)

where f_Q^c is the Debye–Waller factor at T_c, h_Q is the amplitude of the critical contribution, C is a material-dependent factor and $O(T)$ defines a weak T-dependence. In the case of incoherent scattering this critical behaviour can be seen as an anomalous increase of the root-mean-square displacement:

$$\langle r^2 \rangle^{1/2} = r_c (1 - B\sqrt{T_c - T}), \quad T \leq T_c,$$

(2)

where r_c is the displacement at T_c. Finally, for the shear viscosity, MCT predicts a power-law divergence according to

$$\eta = A \left(\frac{T - T_c}{T_c} \right)^{-\gamma}, \quad T > T_c,$$

(3)

where γ is in the range from 2 to 5.

Results and discussion

Our data with respect to $f_Q(T)$ (not shown here), obtained from the integration of the elastic peak, hint at some discontinuity at 235 ± 10 K, i.e. at $T \sim 1.2 T_g$, but at $T > 240$ K we observed a stronger T-dependence than that predicted from Eq. (1). This is due to quasielastic broadening which has a strong effect on $f_Q(T)$ at higher T ($T > 260$ K) [7]. Therefore, accurate determination of the critical temperature from the T-dependence of f_Q alone is subject to large uncertainty. As an alternative we fitted our experimental data for the root-mean-square displacement (Fig. 1), with Eq. (2) and obtained a critical temperature of about 226 K. The inset to Fig. 1 shows the dependence of r on $(T_c - T)^{0.5}$ using a fixed $T_c = 226$ K. The data points for r follow the predicted square-root behaviour in the T-range studied with the chosen T_c value and the extrapolated value of r at $T = T_c$

(r_c in Eq. (2)) is ~ 0.5 Å, in agreement with the Lindemann criterion [5]. However, it has been pointed out earlier that the statistical relevance of the straight line in the inset to Fig. 1 is not very large [8]. One has to support the choice of T_c by additional reasoning. Therefore, in Fig. 2 we attempt to fit the viscosity data using a fixed $T_c = 226$ K. In doing so we get an exponent

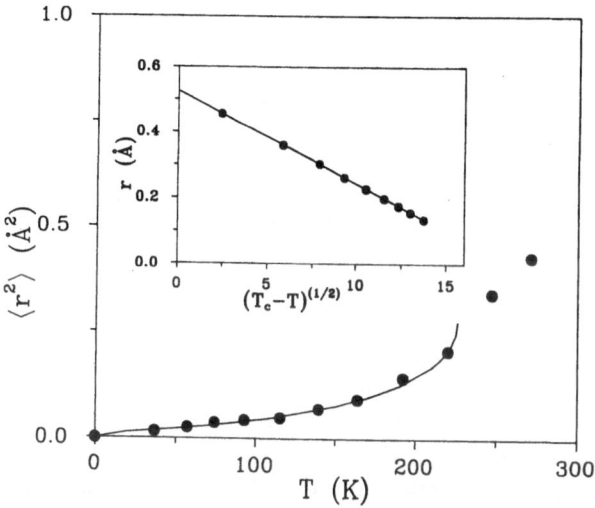

Fig. 1. Temperature dependence of the mean-square displacement. Solid line is the behaviour predicted from the mode coupling theory. In the inset, r is plotted vs. $(T_c - T)^{0.5}$ and displays the predicted (Eq. (2)) linear dependence with $T_c = 226$ K

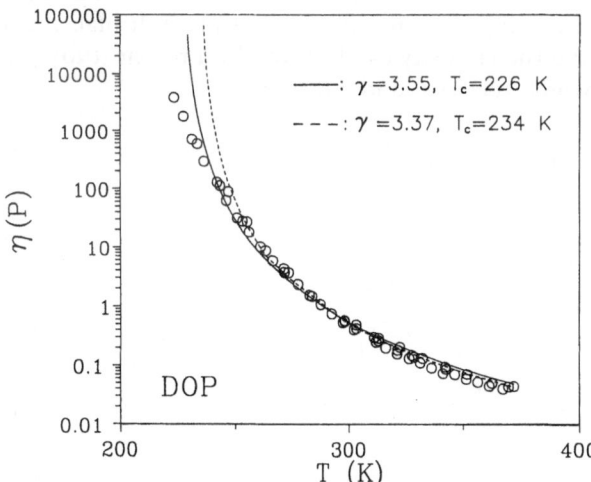

Fig. 2. Temperature dependence of the shear viscosity for DOP. Dashed and solid lines are calculated from Eq. (3) using a fixed T_c value of 234 and 226 K, respectively

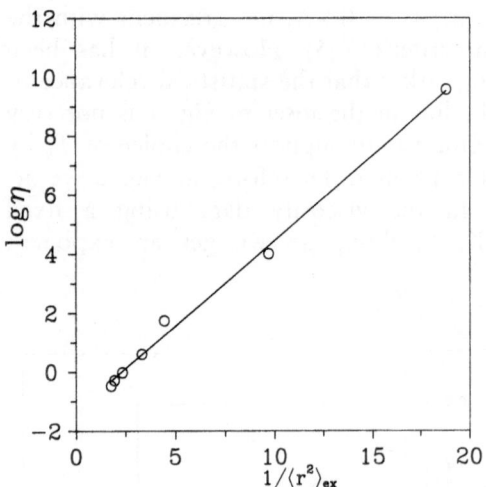

Fig. 3. Relation between the logarithm of the shear viscosity and the inverse of the mean-square-displacement for different temperatures

$\gamma = 3.55$ which is within the predicted range. However, deviations from the experimental data occur especially at high T and a different T_c value, i.e., $T_c = 234$ K can describe the viscosity data better (Fig. 2). Therefore, our data for $f_Q(T)$, r and η obtained from Eqs. (1)–(3), respectively, are internally consistent, but neither of the three can provide T_c unambiguously. We need to mention, however, that the breakdown of the SED equation – from the DRS and DR results [7] – occurred at T near the neutron scattering T_c.

An alternative interpretation to our neutron-scattering data for the mean-square-displacement and the viscosity can be given by a recent study [9], which relates η and r through

$$\eta = \eta_0 \exp\left(\frac{r_0^2}{\langle r^2 \rangle}\right), \tag{4}$$

where r_0 is a critical displacement necessary for a structural rearrangement (α-relaxation). In Fig. 3 we plot the logarithm of the measured viscosity vs. the inverse of $\langle r^2 \rangle_{ex}$, where the latter is calculated by subtracting the low-T Debye–Waller factor due to vibrations [10]. The fit to Eq. (4) shows a linear relation with no discontinuity at T_c, with parameters $r_0 = 1.15$ Å and $\eta_0 = 4.7 \cdot 10^{-2}$ Pa s. It is noteworthy that this approach is based on free-volume arguments and covers a broader T-range.

References

1. Leutheusser E (1984) Phys Rev A 29:2765; Bengtzelius U, Götze W, Sjölander A (1984) J Phys C 17:5915; Götze W (1986) Phys Scri 34:66
2. Petry W, Bartsch E, Fujara F, Kiebel M, Sillescu H, Farago B (1991) J Phys Condensed Matter 83:175
3. Fujara F, Petry W (1987) Europhys Lett 4:921
4. Börjesson L, Elmroth M, Torell LM (1990) Chem Phys 149:209
5. Frick B, Farago B, Richter D (1990) Phys Rev Lett 64:2921; Frick B, Richter D, Petry W, Buchenau U (1988) J Phys Condensed Matter 70:73
6. Mezei F, Knaak W, Farago B (1987) Phys Rev Lett 58:571
7. Floudas G, Higgins JS, Fytas G (1992) J Chem Phys 96:7672
8. Götze W, private communication
9. Buchenau U, Zorn R (1992) Europhys Lett 18:523
10. Floudas G, Higgins JS, Kremer F, Fischer EW (1992) Macromolecules, in press

Received December 16, 1992;
accepted April 30, 1992

Authors' address:

George Floudas
Max-Planck-Institut für Polymerforschung
Postfach 3148
D-W-6500 Mainz, FRG

Progress in Colloid & Polymer Science   Progr Colloid Polym Sci 91:31–34 (1993)

Tracer diffusion in polymer and organic liquids close to the glass transition studied by forced Rayleigh scattering

M. Lohfink and H. Sillescu

Institut für Physikalische Chemie der Johannes Gutenberg-Universität Mainz, FRG

Abstract: Translational diffusion coefficients D of photochromic dye molecules have been measured by forced Rayleigh scattering in polymer diluent systems at compositions from pure polymer to pure diluent. D could be measured down to below the glass transition temperature T_g. In the pure diluent, $D(T_g)$ was found typically about two decades below that in the pure polymer but still far above the Stokes–Einstein prediction extrapolated from high temperatures. Some experimental results are in contrast with predictions from free-volume theory.

Key words: Diffusion – polymers – plasticisers – glass transition

The mechanical properties of polymer glasses can be varied considerably by adding diluent molecules which serve as plasticizers or antiplasticizers depending upon their mobility and packing within the polymer [1–3]. The mobility has been studied extensively by permeation and sorption techniques, where diffusion coefficients D typically in the range 10^{-5}–10^{-10} cm^2 s^{-1} can be measured [4]. By application of the forced Rayleigh scattering (or holographic grating) technique the diffusion of photochromic probe molecules can be followed to much smaller D values down to almost 10^{-17} cm^2 s^{-1} [5]. This has allowed to extend the probe size for diffusion studies to values above the monomer size of the polymer matrix and to investigate diffusion in polymer diluent systems, where the probe size is comparable with that of the diluent [5–9] in the vicinity of the glass transition temperature T_g. An important result of these studies is that varying the probe size by a factor of ~ 10 from rare gas atoms to large organic dye molecules leads to a change of D by about nine decades from 10^{-5} to 10^{-14} cm^2 s^{-1} in typical polymer glasses at T_g [4, 5].

Probe diffusion in polymers and polymer diluent systems has been described in the framework of the free-volume concept by Fujita [10] and was elaborated further in a series of papers by Vrentas, Duda, and their coworkers [11–20]. In the limit of low diluent concentration their result can be written in the form of the Vogel–Fulcher–Tamman (VFT) relation or the equivalent Williams–Landel–Ferry (WLF) equation [21] which is chosen in the present paper. Thus, the solute diffusion coefficient $D(T)$ relative to its value at T_g can be formulated as [21]

$$-\log a_T = \log\left[D(T)/D(T_g)\right]$$
$$= \frac{C'_{1g}(T - T_g)}{C'_{2g} + T - T_g}. \tag{1}$$

C'_{1g} and C'_{2g} are related to the corresponding WLF parameters of the matrix via coupling parameters ξ and λ by

$$C'_{1g} = \xi C_{1g}, \quad C'_{2g} = C_{2g}/\lambda, \tag{2}$$

where $\lambda = 1$ for $T > T_g$, but $\neq 1$ at $T < T_g$. The free-volume interpretation relates C_{1g} and C_{2g} to the "fractional free volume" $f = v_f/v_m$ defined as the ratio of the average free volume v_f over some molecular volume. f is assumed to vary with temperature as $f = f_g + \alpha_f(T - T_g)$ where α_f is the difference of the expansion coefficients above and below T_g and f_g the fractional free volume at T_g. One obtains [21]

$$C_{1g} = B/(f_g \ln 10), \quad C_{2g} = f_g/\alpha_f, \tag{3}$$

where the "size parameter" B is assumed to be equal to one, for simplicity [21]. In polymer diluent systems of arbitrary composition, the application of free-volume theory as developed by Vrentas and Duda implies problems which require additional assumptions. For example, one has to avoid that a negative polymer free-volume contribution arises if the diluent T_g is far below that of the polymer [9].

We have circumvented these problems by chosing an example where T_g is constant over the whole composition range from pure polymer to pure diluent. It is then appropriate to simply replace f by

$$f_{mix} = \Phi_1 f_1 + \Phi_2 f_2 \qquad (4)$$

where Φ_i and f_i are the volume fraction and fractional free volume of components 1 and 2, respectively. The diffusion probe molecules present as a third component are treated in the dilute solution limit by Eq. (1–3) since their concentration is not larger than 0.5%. It is apparent from Eq. (1) that the difference between the temperature dependence of probe diffusion and mechanical relaxation of the matrix can be accounted for by a single temperature independent coupling parameter ξ for $T > T_g$, where $\lambda = 1$. In free-volume treatments, ξ is interpreted as the ratio of the critical molecular volume v_s^* of the solute jumping unit over the critical molecular volume v_m^* of the matrix jumping unit [11–18]. Thus, $\xi = 1$ for self-diffusion in a pure liquid glass former, since $v_s^* = v_m^*$ in this case. However, we have shown in recent experiments that self-diffusion in pure orthoterphenyl (measured by a NMR–FG technique) can be described by $\xi = 1$ only at temperatures $T \gtrsim T_c$, where T_c is about 50 K above T_g, but $\xi < 1$ for $T < T_c$ [22]. This is in constrast to the assumption $\xi = v_s^*/v_m^*$ and renders questionable the whole free-volume treatment of probe diffusion. An alternative picture of cooperative molecular motion at $T_g < T < T_c$ has been proposed which also accounts for the equality of the mean correlation times of structural relaxation and molecular reorientation by the assumption that the lifetime of molecules in cooperatively reorienting aggregates is of the order of the rotational correlation time [23].

However, though this picture is consistent with all current experimental results on translational and rotational diffusion in supercooled liquids of non-spherical molecules, and is in harmony with a "tear and repair mechanism" proposed recently

by Stillinger [24], it has no solid theoretical basis where cooperative molecular motion should be a result rather than an assumption. We refer to ref. [23] for a more extensive discussion of probe diffusion in supercooled liquids, where also results on polymer glass formers are briefly discussed. In the present paper, we concentrate on diffusion in polymer diluent systems as compared with that in the pure diluent. Since the free-volume concept will remain of heuristic value in the near future, in particular, for the dynamics of polymer glass formers, we apply it although we are aware of its inconsistencies.

We have studied the diffusion of the photochromic dye 2,2'-bis(4,4-dimethylthiolan-3-one), abbreviated as TTI [5], in mixtures of polymethylphenyl-siloxane) (PMPS) and bis (3-methoxyphenyl)cyclohexane (BMC) both having $T_g = 243$ K. The mixtures were prepared by dissolution of the components in tetrahydrofuran, filtration through a 0.2 μm Millipore filter, evaporation of the solvent and annealing for 48 h at 430 K under vacuum. The TTI concentration was between 0.3 and 0.5 wt%. We have checked by DSC that T_g was constant (242.5 $\pm$ 1.0 K) for all mixtures, and by index of refraction measurements that there is no volume change on mixing PMPS and BMC [25]. The forced Rayleigh scattering measurements were done as described elsewhere [5, 26] using a probe cell that can be cooled to $T \gtrsim 200$ K by the cooling fluid of a cryostat [25]. In Fig. 1 some typical results of the diffusion measurements are shown with fits by the WLF equation (1). In Fig. 2 the parameters of all

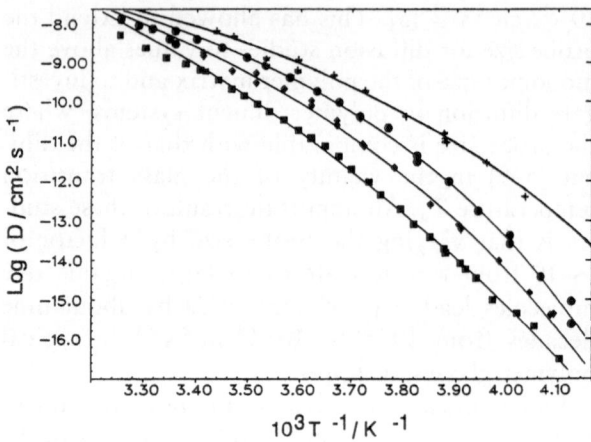

Fig. 1. Diffusion coefficients of the photochromic dye TTI in 0, 3.2, 52 and 100 wt% (from above) BMC in PMPS

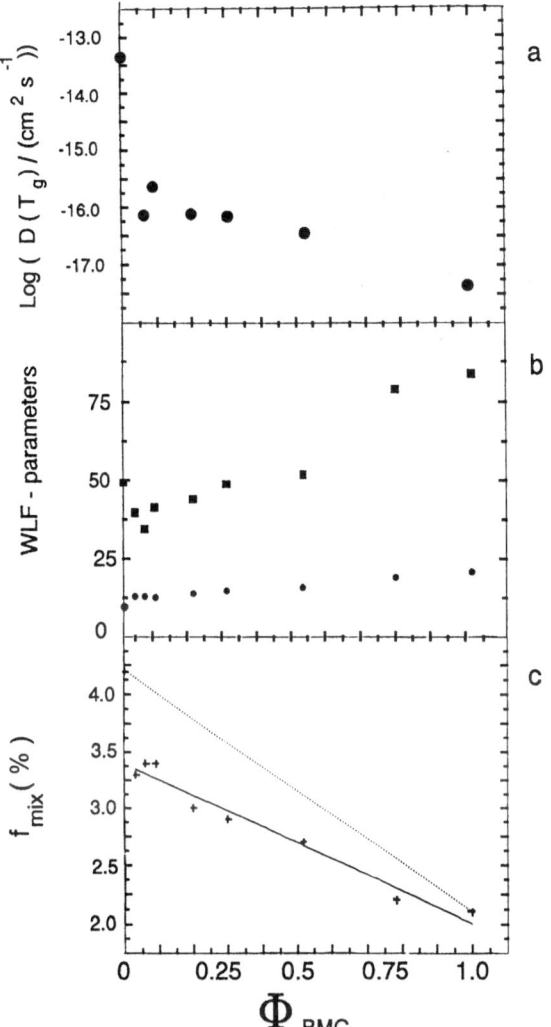

Fig. 2. a) Diffusion coefficients D of TTI in BMC/PMPS at $T = T_g$ drawn versus the BMC volume fraction Φ_{BMC}. b) WLF parameter C'_{1g} (●) and C'_{2g} (■, in K) for TTI diffusion in BMC/PMPS versus Φ_{BMC}. c) Fractional free-volume f_{mix} for TTI diffusion in BMC/PMPS vs. Φ_{BMS} (see text for full and dashed lines)

WLF fits are shown as a function of the BMC volume fraction. The fractional free-volume f_{mix} "seen" by the diffusing probe in the mixtures at T_g was determined from Eq. (3) as $f_{mix} = (C'_{1g} \ln 10)^{-1}$ (symbols in Fig. 2c). The dashed line was obtained from Eq. (4), where f_1 and f_2 are determined at T_g in pure PMPS and BMC, respectively. The full line is a fit omitting the value in pure PMPS. By comparison of the C'_{1g} values determined from the D values in PMPS and BMC with the C_{1g} values determined from fits

of the shear viscosity [25, 27], assuming $C_{2g} = C'_{2g}$ [5], we obtained the coupling parameter $\xi = C'_{1g}/C_{1g}$ as 0.70 and 0.87 in pure PMPS and BMC, respectively.

The main results shown in Fig. 1 are that addition of a few percent of diluent causes a dramatic reduction of $D(T_g)$, and that the curvature of the full lines determined by C_{2g} (see Fig. 2b) is largest for 3.2% diluent concentration, but smallest for the pure diluent. The latter has been found previously in other supercooled liquids where Arrhenius fits $(C_{2g} = T_g)$ have been proposed in some temperature range above T_g [28]. The large drop of $D(T_g)$ on addition of little solvent (see Fig. 2a) indicates that a small fraction of the sorbed molecules fill special sites within the polymer thus "blocking" diffusion over large distances [4]. This effect seems to be most pronounced in our example where T_g is the same for polymer and diluent, whereas the high diluent mobility in "plasticized" polymer glasses can be related with the lower T_g of the diluent in these systems [5–9, 23]. The sharp reduction of f_{mix} seen in Fig. 2c for small Φ_{BMC} which is in contrast to the linear relation given by Eq. (4) (dashed line) is in harmony with the picture of few blocking sites. However, more experiments are necessary to study this effect which should critically depend upon the presence of lower molecular weight impurities (monomers and oligomers) and the dye concentration [7].

Acknowledgement

Support by the Deutsche Forschungsgemeinschaft (SFB 262) is gratefully acknowledged.

References

1. Petrie SEB, Moore RS, Frick JR (1972) J Appl Phys 43:4318
2. Jackson WJ, Caldwell JR (1967) J Appl Polym Sci 11:211, 227
3. Fischer EW, Hellmann GP, Spiess HW, Hörth FJ, Ecarius U, Wehrle M (1985) Makromol Chem Suppl 12:189
4. Crank J, Park GS (1968) Diffusion in Polymers. Academic Press, London
5. Ehlich D, Sillescu H (1990) Macromolecules 23:1600
6. Zhang J, Wang CH (1988) Macromolecules 21:1811
7. Wang CH, Xia JL, Yu L (1991) Macromolecules 24:3638
8. Lodge TP, Lee JA, Frick TS (1990) J Pol Sci Phys 28:2607
9. Frick TS, Huang WJ, Tirrell M, Lodge TP (1990) J Pol Sci Phys 28:2629
10. Fujita H (1961) Fortschr Hochpol Forsch 3:1
11. Vrentas JS, Duda JL (1971) J Appl Polym Sci 21:1215

12. Vrentas JS, Duda JL (1978) J Appl Polym Sci 22:2325
13. Vrentas JS, Liu HT, Duda JL (1980) J Appl Polym Sci 25:1293
14. Vrentas JS, Liu HT, Duda JL (1980) J Appl Polym Sci 25:1297
15. Vrentas JS, Duda JL, Ling HC (1984) J Polym Sci Phys 22:459
16. Vrentas JS, Duda JL, Ling HC (1985) J Polym Sci Phys 23:275
17. Vrentas JS, Duda JL, Ling HC, Hou AC (1985) J Polym Sci Phys 23:289
18. Vrentas JS, Duda JL, Hou AC (1985) J Polym Sci Phys 23:2469
19. Vrentas JS, Duda JL, Hou AC (1986) J Appl Polym Sci 31:139
20. Vrentas JS, Chu CH (1989) J Polym Sci Phys 27:1179
21. Ferry JD (1980) Viscoelastic Properties of Polymers, 3rd ed. Wiley, New York
22. Lohfink M, Fujara F, Sillescu H, Fleischer G (1991) unpublished preprint; Fujara F, Geil B, Hartmann K, Hinze G, Sillescu H, Fleischer G (1992) Z Phys B – Condensed Matter 88:195
23. Lohfink M, Sillescu H (1992) Proceedings of the 1st Tohwa University International Symposium. Fukuoka, Japan, American Institut of Physics Conference Series 256:30
24. Stillinger FH (1988) J Chem Phys 89:6461
25. Lohfink M (1991) Dissertation, Universität Mainz
26. Sillescu H, Ehlich D (1990) In: Fouassier JP, Rabek JF (eds) Appl of Lasers in Polym Sci and Technology, Vol. III, p. 211. CRC Press, Boca Raton, FL
27. Meier G, Gerharz B, Boese D, Fischer EW (1991) J Chem Phys 94:3050
28. Laughlin WT, Uhlmann DR (1972) J Phys Chem 76:2317

Received January 22, 1992;
accepted June 15, 1992

Authors' address:

H. Sillescu
Inst. f. Physikalische Chemie
Univ. Mainz
Jakob-Welder-Weg 15
D-W-6500 Mainz, FRG

Progress in Colloid & Polymer Science   Progr Colloid Polym Sci 91:35–38 (1993)

Light scattering studies of a glass forming liquid near T_g

A. Patkowski, E. W. Fischer, H. Gläser, G. Meier, H. Nilgens, and W. Steffen

Max-Planck-Institut für Polymerforschung, Mainz, FRG

Abstract: The cluster growth dynamics and the equilibrium cluster correlation length were studied in 1,1-di(4′-methoxy-5′-methyl-phenyl)-cyclohexane (BMMPC also called BKDE in previous papers) in the temperature range from 40 °C to 70 °C by means of static light scattering (SLS) and photon correlation spectroscopy (PCS). The equilibrium cluster correlation length obtained from PCS and SLS using the Ornstein Zernike formula are in a good agreement with each other. The analysis of the SLS data based on the Debye formula results in much lower values of the cluster correlation length. The relaxation processes in BMMPC were studied using depolarized Rayleigh spectroscopy (DRS). The relaxation times obtained from DRS exhibit an Arrhenius behaviour and with decreasing temperature diverge from the relaxation times of the primary α process measured by means of dielectric spectroscopy which follow the usual Williams–Landel–Ferry temperature dependence.

Key words: Static, dynamic correlation length – static, dynamic light scattering – equilibrium cluster size – dynamics of glass forming liquids

Introduction

As it has been established recently [1], low molecular weight and polymeric glass forming liquids exhibit unusual light-scattering characteristics not predicted by theories, i.e. excess isotropic Rayleigh intensity, R_{iso}^{exc}, additional slow component in the polarized photon correlation function, and high Landau Placzek ratio, R_{LP} [1–7]. Their unusual features have been explained by long-range density fluctuations – clusters, which can be characterized by "static" and "dynamic" correlation length obtained from the angular dependence of the $R_{iso}^{exc}(q)$ and from the polarized correlation function, respectively [2, 4, 6]. The experimental results obtained so far for several glass forming liquids are qualitatively in a good agreement; however, the data are not quantitatively reproducible because the values of a correlation length at a given temperature depend on the thermal history of the sample [4, 6]. In order to solve this problem, we have performed a light-scattering study on BMMPC and attempted to answer the following questions: (i) is

there a characteristic equilibration time for cluster growth and annealing for a given temperature, (ii) can a sample after reaching an equilibrium be characterized by reproducible values of the "static" and "dynamic" correlation length, and (iii) what is the temperature dependence of the equilibrium correlation length and the equilibration times. In order to compare the collective dynamics measured by depolarized Rayleigh spectroscopy with results obtained previously by means of dielectric relaxation [8] the depolarized Rayleigh spectra of BMMPC were measured at high temperatures above T_g. With decreasing temperature an increasing deviation of the DRS correlation times from the α process measured previously was observed.

Experimental

The static light scattering (SLS), i.e. the measurements of the angular dependence of the polarized and depolarized scattered intensity, were performed using an experimental system described previously [3, 4, 7]. The incident beam from a krypton-ion laser ($\lambda = 647$ nm) was polarized vertically. Vertical (polarized) and horizontal (depolarized) components of the scattered light were measured for scattering angles, θ, from $25°$ to $140°$. Thus, the accessible range of the scattering vector $q = (4\pi n/\lambda)\sin(\theta/2)$ amounted to 0.008–0.03 nm^{-1}. In order to diminish the effects of the non-ergodicity of the sample at low temperatures the sample cell was rotated at frequencies from 0.5 to 5.5 rpm. The photon correlation functions were measured using Fabry-Perot interferometers (FPI) of a free spectral range (FSR) amounting to 7.5 GHz and 750 MHz. The incident beam was supplied by an argon-ion laser ($\lambda = 514$ nm). The synthesis and purification of BMMPC is described elsewhere [5]. Light-scattering samples were prepared by filtering them three times through a 0.22 μm millipore filter in dust-free pyrex cells, made out of cylindrical tubes. The whole filtering apparatus was maintained at 80 °C during filtering.

Results and discussion

An apparent non-ergodicity of BMMPC at 40 °C in the time scale of the light-scattering experiment, i.e. the fact that the time-average intensity $\langle I \rangle_T$ is not equal to the ensemble-average $\langle I \rangle_E$, is clearly demonstrated in Fig. 1. One set of the SLS data was obtained for a fixed sample, each point is an average of 20 measurements 20 s each, and this is the time-average data. The ensemble-average intensities were measured by rotating the sample at 5.5 rpm and each point is an average of three measurements 3 s each. The vertical bars in Fig. 1 represent the fluctuation of the data which was much higher for a fixed sample. The shape of the SLS curve measured for a fixed sample was not reproducible and it represents fluctuations of speckles. All the static data discussed in this paper were measured with a rotating sample except the data above 55 °C, where the probe rotator could not be used any more. The isotropic Rayleigh ratio $R_{iso}(q)$ can be calculated from the polarized $R_{VV}(q)$ and depolarized $R_{VH}(q)$ Rayleigh components and consists of $R'_{iso}(q)$ which is entirely due to the isothermal compressibility and an excess contribution $R_{iso}^{exc}(q)$ due to clusters: $R_{iso}(q) = R'_{iso}(q) + R_{iso}^{exc}(q) = R_{VV}(q) - \frac{4}{3}R_{VH}(q)$. The static correla-

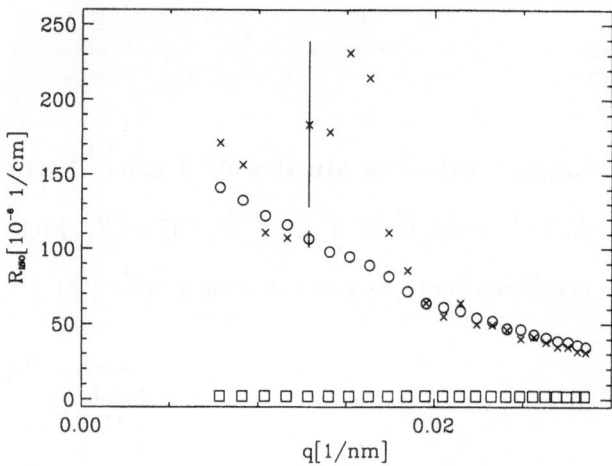

Fig. 1. The isotropic Rayleigh component $R_{iso}(q)$ measured with a fixed sample cell ($\times$) and rotating sample cell ($\bigcirc$) and the anisotropic Rayleigh component $R_{VH}(q)$ ($\square$) for BMMPC at $40°$

tion length can be calculated using either the Debye formula:

$$R_{iso}^{exc}(q) = \frac{8\pi^3}{\lambda^4}\frac{\langle \delta\varepsilon^2 \rangle \xi_D^3}{(1 + \xi_D^2 q^2)^2}, \tag{1}$$

or the Ornstein Zernike formula:

$$R_{iso}(q) = R_{iso}(q = 0)\frac{1}{1 + \xi_O^2 q^2}. \tag{2}$$

This is achieved by plotting $1/\sqrt{R_{exc}^{iso}(q)}$ (Debye) or $1/R_{iso}(q)$ (Ornstein Zernike) versus q^2. From the linear regression one obtains the slope and the intercept and, thus, the correlation length $\xi = \sqrt{\text{slope/intercept}}$. In these formulas, $\langle \delta\varepsilon^2 \rangle$ is the mean square fluctuation of the dielectric constant, λ is the wavelength of the incident light, and ξ_D and ξ_D are the Debye and Ornstein Zernike "static" correlation lengths, respectively. In order to be able to obtain quantitative parameters characterizing the excess isotropic Rayleigh component and the additional slow component in the VV correlation function, i.e. the static and dynamic correlation length, the experimental data have been analysed using the theoretical basis of critical phenomena [9, 10]. A detailed description of the data analysis is given elsewhere [4, 6, 7]. The cluster contribution to the measured autocorrelation function of the polarized scattered light is a single exponential correlation function and can be expressed as

$$G^{(2)}(\tau) = A(1 + B|g^{(1)}(\tau)|^2) , \qquad (3)$$

where $g^{(1)}(\tau) = \exp(-\Gamma\tau)$, A is a baseline, and B is a constant. The linewidth $\Gamma = 1/\tau_c = Dq^2$, where τ_c is the correlation time of cluster dynamics and D is a translational diffusion coefficient of the clusters. The dynamic correlation length can be calculated as

$$\xi_{dyn} = \frac{kT}{6\pi\eta D} , \qquad (4)$$

where k is the Boltzmann constant and η is the viscosity. The equilibration times of cluster growth were measured in the following way: First the BMMPC sample was equilibrated at 70 °C until all the parameters measured in light scattering were constant. Then the sample was put into the light-scattering instrument in which the desired temperature was already achieved (temperature jump). The polarized, VV, intensity at an angle of $\theta = 40°$ was monitored as a function of time. In the mean-time, the complete SLS data and the correlation functions were measured. From the exponential increase of the scattered intensity at $\theta = 40°$, the equilibration times were calculated. After this intensity reached the plateau value the equilibrium static and dynamic correlation lengths were measured. The equilibration times of 4.3, 18.6, 52.9,

and 156.3 h were obtained for temperature jumps from 70 °C to 60 °C, 50 °C, 45 °C and 40 °C, respectively. The equilibrium correlation times of cluster dynamics τ_c measured for equilibrated and non-equilibrated samples as well as equilibration times at different temperatures are shown in Fig. 2. As one can see the τ_c^n values for the non-equilibrated samples are lower, they approach τ_c^e values of the equilibrated sample with equilibration time. The equilibrium values of the static and dynamic correlation lengths at different temperatures are shown in Fig. 3. A reasonable agreement of the static ξ_O [Eq. (2)] and the dynamic ξ_{dyn} [Eq. (4)] values can be seen, while the ξ_D value [Eq. (1)] is much lower. The substantial scattering of the dynamic ξ_{dyn} values measured at different scattering angles is probably due to the non-ergodicity of the sample. These values should be calculated form the ensemble-average and not from the time-average correlation function. A systematic analysis of the light-scattering data based on the theoretical predictions for non-ergodic systems [11] is currently under way in our laboratory. In Fig. 2, the relaxation times obtained in this study by means of depolarized Rayleigh spectroscopy [13] are compared with the relaxation times measured previously by means of dielectric spectroscopy (DS) [8]. At the highest temperatures, the DRS relaxation times are in a good agreement with the DS

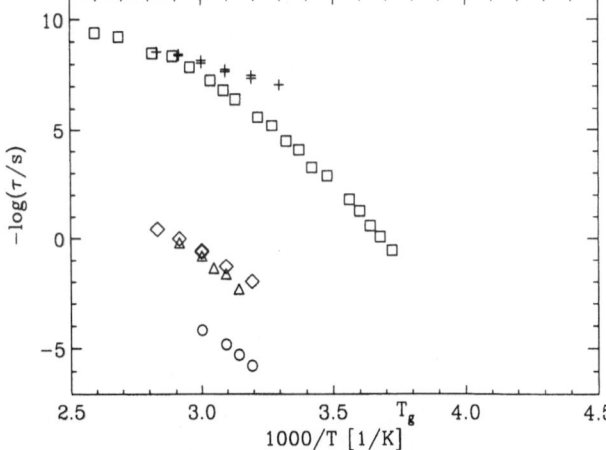

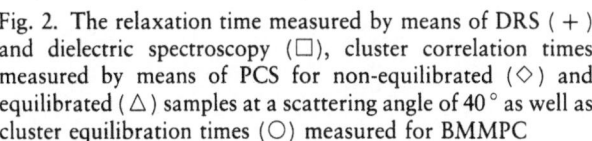

Fig. 2. The relaxation time measured by means of DRS (+) and dielectric spectroscopy (□), cluster correlation times measured by means of PCS for non-equilibrated (◇) and equilibrated (△) samples at a scattering angle of 40° as well as cluster equilibration times (○) measured for BMMPC

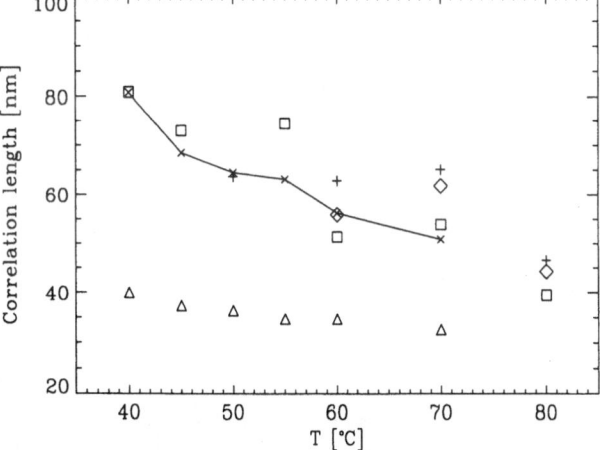

Fig. 3. Correlation length for BMMPC measured by means of SLS and calculated using Debye (△) and Ornstein–Zernike (×) formulas as well as measured by means of PCS at different scattering angles: (□) $q = 0.028$ nm^{-1}; (◇) $q = 0.021$ nm^{-1}; and (+) $q = 0.010$ nm^{-1}

data, which describe the primary α process. However, with decreasing temperature there is an increasing difference between the DRS and DS relaxation times, i.e. DRS relaxation times follow the Arrhenius temperature dependence, while the α process measured by means of DS exhibits the usual Williams–Landel–Ferry behaviour. Previously, the relaxation times measured by means of DRS for *o*-terphenyl [12] were found to be in a good agreement with DS and NMR data, all of them describing the primary α process. In this respect, the discrepancy between the DRS and DS relaxation times is a qualitatively new effect which may be due to bifurcation of the single dynamic process observed at high temperatures into the primary α process and secondary β process. Similar behaviour has been observed in our laboratory for another glass forming liquid, 1,1-di(4'-methoxyphenyl)-cyclohexane (BMPC). In order to assign the DRS relaxation process properly, low-temperature data below T_g are necessary. Such measurements are currently under way in our laboratory.

Acknowledgement

Partial support of the Deutsche Forschungsgemeinschaft (SFB 262) is gratefully acknowledged.

References

1. Fischer EW (1990) In: Colmenero J, Alegria A (eds) Basic Features of the Glassy State. World Scientific, Singapore, New Jersey, London, Hong Kong, p 172
2. Fischer EW, Becker C, Hagenah J, Meier G (1989) Prog Colloid Polym Sci 80:198
3. Fischer EW, Meier G, Rabenau T, Patkowski A, Steffen W, Thoennes W (1991) J Non Cryst Solids 131
4. Rayleigh and Rayleigh-Brillouin spectroscopic study of ortho-terphenyl above the glass transition (to be published)
5. Gerharz B, Meier G, Fischer EW (1990) J Chem Phys 92:7110
6. Thoennes W (1990) Ph.D. Thesis, Mainz
7. Baumann M (1990) M.S. Thesis, Mainz
8. Meier G, Gerharz B, Boese D, Fischer EW (1991) J Chem Phys 94:3050
9. Swinney HL (1974) In: Cummins HZ, Pike ER (eds) Photon Correlation and Light Beating Spektroscopy. Plenum Press, New York, London, p 331
10. Stepanek P, Lodge TP, Kedrowski C, Bates FS (1991) J Chem Phys 94:8289
11. Pusey P, van Megen W (1989) Phys A 157:705
12. Steffen W, Patkowski A, Meier G, Fischer EW (1992) J Chem Phys 96:4171
13. Nilgens H (1991) M.S. Thesis, Mainz

Authors' address:

Adam Patkowski
MPI für Polymerforschung
Ackermannweg 10
W-6500 Mainz, FRG

Progress in Colloid & Polymer Science   Progr Colloid Polym Sci 91:39–42 (1993)

The scaling of the α-relaxation in polymers and low-molecular glass-forming liquids – a comparison

A. Schönhals, F. Kremer[1]), and E. Schlosser

Zentrum für Makromolekulare Chemie, Berlin, FRG
[1]) Max-Planck Institut für Polymerforschung, Mainz, FRG

Abstract: By measuring the complex dielectric susceptibility over 15 decades, we compare the scaling behavior of the α-relaxation for low-molecular-weight glass-forming liquids and polymers. The characteristic differences found for these both classes of substances are discussed in terms of segmental dynamics.

Key words: α-relaxation – dielectric relaxation – nonexponential behavior

Introduction

The dynamical behavior of supercooled liquids or polymer melts, known as α-relaxation, is quite complex. Besides the rapid increase of the characteristic relaxation time observed when a sample is cooled to the glass transition, the nonexponential (or non-Debye) shape of the relaxation function is one of the most important features of the α-relaxation [1]. From a theoretical point of view it seems to be clear that this non-Debye behavior is related to the cooperative nature of motions involved in the α-relaxation, but a true microscopic description remains an unsolved problem in condensed matter physics [1]. Experiments in very broad frequency range are necessary to gain more information to solve this question. Moreover, both a comparison of the relaxation behavior of simple and complex systems like monomers and polymers and a study of the temperature dependence of the shape of the relaxation function in a wide temperature range seems to be useful. By means of broadband dielectric spectroscopy, it is possible to study the α-relaxation from temperatures close to the glass transition temperature T_g up to high temperatures.

Experimental

To study the complex dielectric permittivity

$$\varepsilon^*(f) = \varepsilon'(f) - i\varepsilon''(f) , \qquad (1)$$

(ε' – real part, ε'' – loss part, $i = \sqrt{-1}$) at the α-relaxation over 15 decades of the frequency f, five different measuring systems were employed [2, 3]: a time-domain spectrometer [4] ($10^{-5} < f/\mathrm{Hz} < 1$), a frequency-response analyzer (Schlumberger 1260) with a buffer amplifier of variable gain [4] ($10^{-1} < f/\mathrm{Hz} < 10^6$), an impedance analyzer (HP 4192 A; $10^2 < f/\mathrm{Hz} < 10^7$), a coaxial line reflectometer (HP 4191 A; $10^6 < f/\mathrm{Hz} < 10^9$) and a network analyzer (HP 8510 E; $5 \cdot 10^7 < f/\mathrm{Hz} < 2 \cdot 10^{10}$).

In order to compare directly the relaxation behavior of low-molecular-weight glass-forming liquids and polymers propylene glycol (PG) and its polymeric counterpart poly(propylene glycol) (PPG) were chosen. In addition, poly(vinyl acetate) (PVAC) and poly(p-chlorostyrene) (PCLS) which are typical amorphous polymeric substances with quite different structures were selected.

ε'' measured for propylene glycol in the available frequency range is shown in Fig. 1. As the temperature is lowered, the peak frequency f_p of ε'' decreases (see Fig. 2). A similar behavior is found for the polymeric systems (Fig. 2). The well-known Vogel–Fulcher equation [6]

$$\log f_p = \log f_{p\infty} - A/(T - T_0) , \qquad (2)$$

(A and $f_{p\infty}$ are constants; T_0 is the so-called Vogel temperature), often used to parametrize $f_p(T)$, fits the data only in the low temperature range

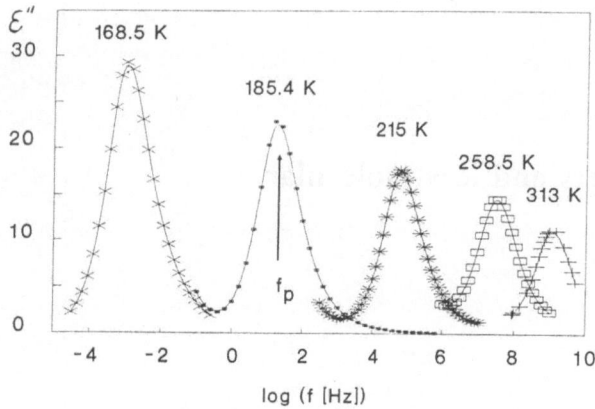

Fig. 1. $\varepsilon''(f)$ of propylene glycol at the indicated temperatures. Each curve represents an example for the measuring system used

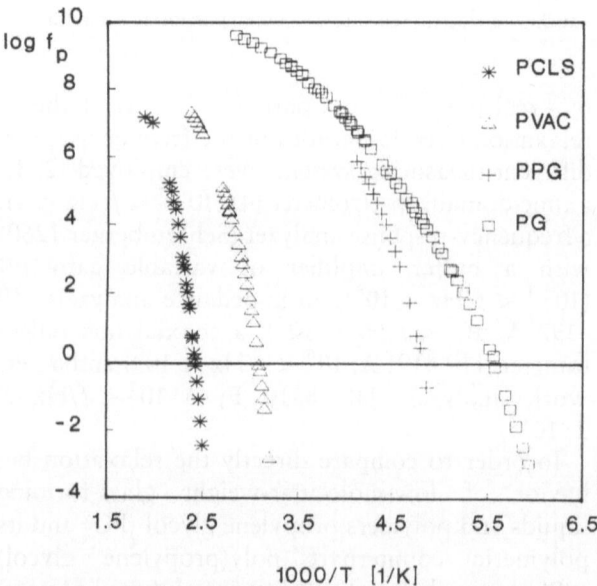

Fig. 2. $\log f_p$ vs. $1/T$ for the different samples. In the frequency range of our experiment PPG shows a normal-mode process [7] at low and secondary relaxations [7] at high temperatures in addition to the α-process. For these reasons the study of the α-relaxation of PPG is restricted from 10^{-1} Hz to 10^6 Hz

($f < 10^8$ Hz). This is discussed in more detail elsewhere [8].

It is well known that the frequency dependence of $\varepsilon^*(f)$ near the α-peak can be represented by a four parameter model function introduced by Havriliak and Negami [9] (HN-function) [10]

$$\varepsilon^*(f) - \varepsilon_\infty = \Delta\varepsilon[1 + (i \cdot f \cdot 2 \cdot \pi \cdot \tau)^\beta]^{-\gamma}$$

$$(0 < \beta, \beta \cdot \gamma \leq 1) . \qquad (3)$$

$\Delta\varepsilon$ and τ denote the intensity and the characteristic relaxation time of the α-process, respectively, and β and γ characterize the shape of the relaxation function. ε_∞ is the limit of $\varepsilon'(f)$ for $f \gg \tau^{-1}$. With regard to the shape parameters, it is convenient to discuss the limiting behavior of $\varepsilon''(f)$ at low and high frequencies which is given by power or scaling laws

$$\varepsilon''(f) \sim f^m \quad (f \ll \tau^{-1}) ;$$
$$\varepsilon''(f) \sim f^{-n} \quad (f \gg \tau^{-1}) . \qquad (4)$$

The scaling parameters m and n $(0 < m, n \leq 1)$ are related to the shape parameters β and γ by the simple formulas $m = \beta$ and $n = \beta \cdot \gamma$, where a simple Debye relaxation is given for $m = n = 1$.

All parameters can be determined by fitting Eq. (3) to the experimental values. Using a general evaluation strategy [10], firstly, the influence of neighboring relaxation regions can be taken into consideration and, secondly, a consistent evaluation of measurements carried out in the frequency and/or time domain is possible.

Results and discussion

The main results of our study are presented in Figs. 3 and 4, where m and n are plotted versus $\log f_p$ so that all samples have the same representation irrespective of their T_g. In Fig. 3, the scaling parameters are compared for PG and PPG. Evidently, the temperature dependence of m and n is completely different for the polymer and the monomer. For PG as a low-molecular-weight glass-forming liquid, $m = 1$ is measured over the whole temperature range and n is significantly greater than 0.5. At high temperatures, n approaches to 1. This means that the shape of the relaxation function is Debye-like at high temperatures in accordance with Dixon et al. [11]. For this reason, in low-molecular systems, the elementary mechanism of motion is an isotropic rotation of an isolated dipole. For PPG, $m = 0.9$ and $n = 0.4 < 0.5$ is measured over the whole range of temperatures. The other two polymers, PVAC and PCLS (Fig. 4) show nearly the same behavior. The scaling parameter m is smaller than 1 and approaches 1 at high temperatures. The high-frequency parameter n depends on temperature at low temperatures and reaches a plateau value of 0.5 at

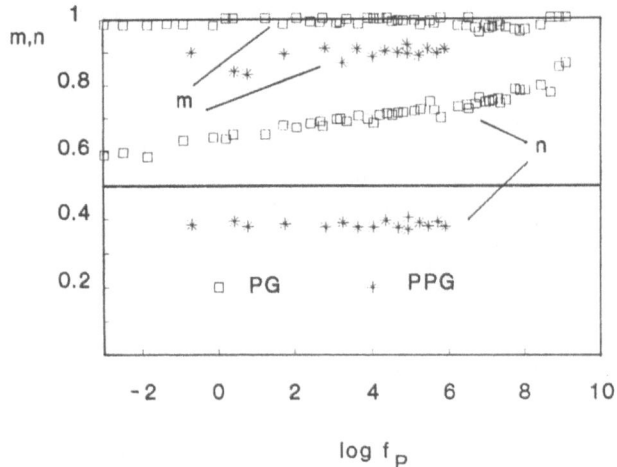

Fig. 3. Scaling parameters m and n vs. $\log f_p$ for PG and PPG

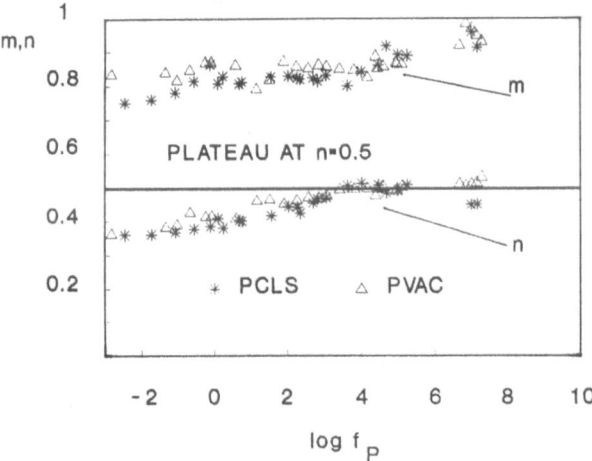

Fig. 4. Scaling parameters m and n vs. $\log f_p$ for PCLS and PVAC

high temperatures. Therefore, it has to be concluded that the shape of the relaxation function shows a non-Debye behavior for all measured temperatures and that the elementary mechanism of motion cannot be an isotropic rotation of isolated dipoles. The bond motions in a polymer chain are much more complicated than the Brownian motion of a small molecule. The connectivity of the chain, which implies that no bond can be displaced independently of its neighbors, has to be taken into account. Helfand [12] and Monnerie [13] could show that the chain connectivity (with regard to the bond orientation) leads to a chain dynamics

which corresponds to a damped diffusional propagation of orientation along the chain. Using a simple two-state model [12], the basic dynamic quantity, the autocorrelation function $cf(t)$, for a single chain in dilute solution is given by

$$cf(t) = \exp(-t/\tau_1) \cdot \exp(-t/\tau_2) \cdot I_0(t/\tau_1) , \quad (5)$$

where τ_1 is the characteristic time responsible for the diffusion of orientation along the chain and τ_2 is a time constant due to the ceasing of the diffusion of orientation, because the individual changes are not necessarily coupled. $I_0(x)$ is the modified Bessel function. For $t \gg \tau_1$, $cf(t)$ behaves like

$$cf(t) \sim (t/\tau_1)^{-\nu} \cdot \exp(-t/\tau_2) , \quad (6)$$

with $\nu = 0.5$. For the loss function $\varepsilon''(f)$ the Fourier transform of Eq. (6) gives power laws with $m = 1$ and $n = 0.5$.

In our dielectric experiments, PVAC and PCLS in the bulk state show at high temperatures (with respect to the shape of the loss function) a relaxation behavior ($n = 0.5$ and $m = 1$, cf. Fig. 4) which is characteristic for isolated polymer chains in dilute solution. Thus, it must be concluded that the conformational diffusion process (local chain motion) is the elementary mechanism of motion for the α-relaxation in polymers. Moreover, these results give more evidence for a recently published model [14] for the interpretation of the shape of the dielectric function at the α-relaxation. In this model, the scaling parameter n (limited to $0 < n \le 0.5$) is related to the local chain motion, whereas m ($0 < m \le 1$) is connected with more cooperative modes of motion. In the framework of this model the decrease of n with decreasing temperature can be explained by an increasing influence of the environment to the relaxation of a reference dipole which leads to a stronger hindrance of the local chain motion. This is also in agreement with recent more microscopic calculations [15].

Compared with PVAC and PCLS, PPG shows a slightly different temperature dependence of the scaling parameters ($n = 0.4$, $m = 0.9$) in the whole range of temperatures. In this system, the OH-groups are responsible for strong hydrogen bonds which are stable up to high temperatures compared with the T_g of PPG. Clearly, these bonds will influence the process of local chain motion and, for this reason, the relaxation behavior of isolated chains ($n = 0.5$, $m = 1$) cannot be observed.

Acknowledgements

The financial support of A. Schönhals by the Max-Planck-Society is gratefully acknowledged. We also thank E. W. Fischer for many helpful discussions and A. Hofmann for the dielectric measurements with the network analyzer.

References

1. Ngai KL (ed) (1990) Proceedings of the Conference on Relaxation in Complex Systems. Crete [J Non-Cryst Solids 131–133]
2. Kremer F, Boese D, Meier G, Fischer EW (1989) Prog Colloid Polym Sci 80:129
3. Schönhals A, Kremer F, Schlosser E (1991) Phys Rev Lett 67:999–1002
4. Schlosser E, Schönhals A (1991) Polymer 32:2135–2140
5. Pugh J, Ryan T (1979) IEEE Conf Dielectric Material Measurement Application 177:404
6. Vogel H (1921) Z Phys 22:645; Fulcher GS (1925) J Amer Chem Soc 8:339
7. Johari GP (1986) Polymer 27:866; Baur ME, Stockmayer H (1965) J Chem Phys 43:4319
8. Schönhals A, Kremer F, Hofmann A, Fischer EW, Schlosser E (1992) Phys Rev Lett (submitted)
9. Havriliak S, Negami S (1966) J Polym Sci Polym Symp 14:89
10. Schlosser E, Schönhals A (1989) Colloid Polym Sci 267:963
11. Dixon PK, Wu L, Nagel SR, Williams BD, Carini JP (1990) Phys Rev Lett 65:1108
12. Hall CK, Helfand E (1982) J Chem Phys 77:3275
13. Monnerie L et al. (1987) In: Richter D, Springer T (eds) Polymer Motion in Dense Systems. Springer Proceedings in Physics 20, Springer, Berlin
14. Schönhals A, Schlosser E (1989) Colloid Polym Sci 267:125
15. Bahar I, Erman B, Kremer F, Fischer EW, Macromolecules, in press

Received January 23, 1992;
accepted June 15, 1992

Authors' address:

Dr. Andreas Schönhals
Zentrum für Makromolekulare Chemie
Rudower Chaussee 5,
D-O-1199 Berlin, FRG

Progress in Colloid & Polymer Science   Progr Colloid Polym Sci 91:43–45 (1993)

Scaling behaviour in poly(propylene glycol) in the glass transition range

D. L. Sidebottom, R. Bergman, L. Börjesson, and L. M. Torell

Chalmers University of Technology, Department of Physics, Gothenburg, Sweden

Abstract: We have performed dynamic light-scattering experiments on poly(propylene glycol) (PPG) above and below the glass transition temperature (200 K). In addition to the α-relaxation (stretched exponential decay with stretching parameter $\beta_K \approx 0.40$), we observe two relaxation processes not witnessed before by photon correlation spectroscopy. They are attributed to the α'- and β-relaxations and they exhibit similar power law decays. All three processes are analysed in terms of recent mode coupling theories. The scaling relations inherent in the mode coupling approach are tested by determining the relevant exponents directly from the spectra.

Key words: Glass transition – light scattering – relaxation – scaling behaviour

Much attention is presently devoted to the physical processes underlying the glass transition mainly due to the progress of the mode coupling theories (MCT) [1–3]. As applied to glassy systems, MCT exploits the coupling that exists between the decay of density fluctuations and the viscosity associated with the surrounding medium. This microscopic coupling results in a feedback mechanism which leads to an enhancement of the macroscopic viscosity as density increases. Ultimately it leads to a divergence of the viscosity when the system achieves a critical density or equivalently a critical temperature, T_c, according to

$$\eta = \eta_0 (T - T_c)^{-\gamma} . \tag{1}$$

Two relaxation processes can be distinguished in the MCT density–density correlation function. The fastest (β-relaxation) represents the relaxation of molecules trapped in a cage of neighboring molecules to the non-ergodic level, f, imposed by the cage. The β-relaxation approaches f in the following power law fashion:

$$\phi_\beta(t) = f + h_\beta \left(\frac{t}{\tau_\beta} \right)^{-a} . \tag{2}$$

Above the transition temperature, T_c, collective rearrangements continue to occur allowing for the eventual relaxation of the cages, the α-relaxation process. MCT predicts that the initial part of the α-process obeys the von Schweidler decay law

$$\phi_\alpha(t) = f - h_\alpha \left(\frac{t}{\tau_\alpha} \right)^b , \tag{3}$$

which is essentially a first-order expansion of the Kohlrausch, stretched exponential form

$$\phi(t) = \exp\left[-\left(\frac{t}{\tau} \right)^{\beta_K} \right], \quad 0 < \beta_K \leq 1 . \tag{4}$$

In this paper we present dynamic light-scattering measurements of the relaxation function associated with the decay of density fluctuations in poly(propylene glycol) of molecular weight 4000 (PPG 4000). We observe three relaxation processes which are attributed to the α', α, and β relaxations, respectively. The exponents a and b are obtained directly from the measured relaxation function and are used to test the MCT relations.

Photon correlation spectroscopy (PCS) was performed in homodyne fashion for vertically polarized light scattered at a scattering angle of 90°. Experimental details are given elsewhere [4]. For temperatures from 208 to 215 K the measured correlation function of $\sigma\phi^2(t)$, where $\phi(t)$ is the density autocorrelation function σ is the coherence factor, display the familiar shape commonly associated with the α-relaxation (see Fig. 1). They are highly non-exponential and well-fitted by Eq. (4) using an exponent, β_K, of 0.39 ± 0.03, in agreement with the reported data for the α-process [5, 6]. However, for $T > 223$ K the form of the relaxation curve is changed revealing a long-time tail. This is discussed in more detail in ref. [4]. To investigate this tail in detail the spectra recorded at the highest temperatures are plotted in the double logarithmic plot (Fig. 2a). The adherence of these spectra to the power law of Eq. (2) suggests an interpretation in terms of the β-relaxation. However, as this relaxation extends past that of the α-process, it cannot qualify as the faster β-relaxation that is associated with cages in the MCT picture. A more probable explanation is that the long-time tail is the precursor of the α′-process associated with diffusional motion of single polymer chains in the liquid [6]. Below the glass transition weak signals revealed the presence of a third process (see Fig. 2b) whose dynamics is attributed to the β-process.

Having assigned the three types of dynamics we proceed to compare with MCT predictions. It is convenient to begin the discussion with the β-process, for which the results at 186 and 198 K are displayed in Fig. 2b in a double logarithmic plot. Although the amplitude is small, and the statistical scatter detrimental, the data cluster about a line whose slope is estimated to be $a = 0.23 \pm 0.05$. To obtain the second exponent, we analyse the initial decay of the α-relaxation and assume that the short-time limit of the spectrum is a sufficient ap-

proximation of f. Hence, plots of $\sigma^{1/2}[f - \phi(t)]$ are displayed in Fig. 3. Here, the short-time von Schweidler relaxation is apparent over about the first three decades and the slope of the linear portion corresponds to an exponent $b = 0.36 \pm 0.02$, independent of temperature.

In addition to the power law forms predicted and above verified for the relaxation function, MCT gives additional conditions which a and b must satisfy [3]. The exponents a and b are closely linked and two relations arise from scaling arguments. The first concerns the coupling parameter, λ, which quantifies the strength of the feedback mechanism:

$$\lambda = \frac{\Gamma(1-a)^2}{\Gamma(1-2a)} = \frac{\Gamma(1+b)^2}{\Gamma(1+2b)}, \tag{5}$$

where $\Gamma(x)$ is the gamma function. Equation (5) gives for the measured exponent $a = 0.23$ a value of $\lambda = 0.88$ and in the case of the result $b = 0.36$ we obtain $\lambda = 0.87$. Thus, the present two exponents satisfy Eq. (5) remarkably well. The second scaling

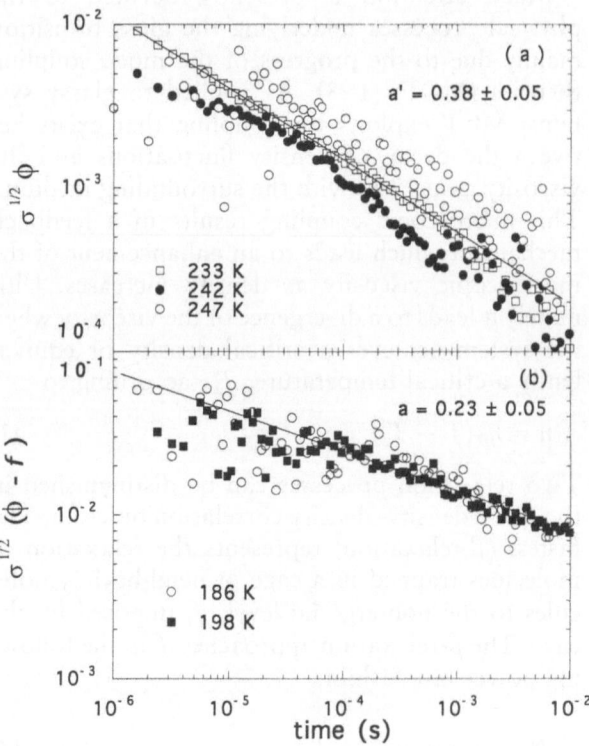

Fig. 2. Log–log plot of the autocorrelation function $\sigma^{1/2}\phi(t)$ for PPG; data obtained at 233, 242, and 247 K; (b) data obtained below T_g at 186 and 198 K, with the nonergodic level subtracted out. Solid lines represent linear fits

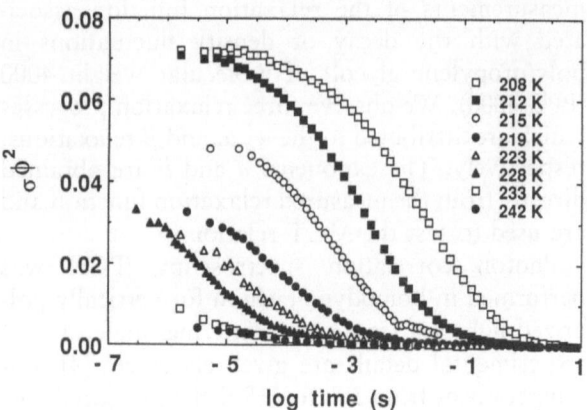

Fig. 1. Autocorrelation function, $\sigma\phi^2(t)$, of PPG 4000 from 208 to 242 K

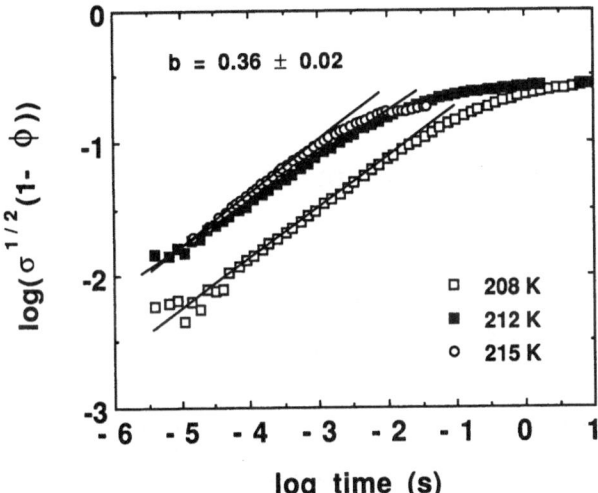

Fig. 3. Plot of $\sigma^{1/2}(1 - \phi(t))$ for spectra collected at 208, 212 and 215 K displaying the short time von Schweidler decay with $b = 0.36 \pm 0.02$

relation concerns the exponent γ in Eq. (1) describing the viscosity divergence,

$$\gamma = \frac{1}{2a} + \frac{1}{2b}. \tag{6}$$

Applying Eq. (6) to the measured exponents yields $\gamma = 3.7 \pm 0.8$. Such an exponent has indeed been found to describe the temperature dependence of the viscosity in a wide temperature range from 260 to 350 K.

Last we analyse the relaxation decay observed at the highest investigated temperatures, which is attributed to the precursor of the α'-process. In a double logarithmic plot (Fig. 2a) we obtain the exponent $a' = 0.38 \pm 0.05$ (the prime refers to the α'-process). Thus, we conclude that the time decay of all processes – β, α and α'– are well described by the functional forms of MCT.

The strength of the present wide-time window PCS measurements is that the exponents a and b are obtained directly from the spectra independent of the MCT relationships. As a consequence

we have been able to perform a stringent test of MCT. The results indicate that the relaxation function is well represented by the power law forms proposed by MCT. Further, the obtained exponents clearly satisfy the scaling relations. The first relation is nontrivial and involves a transcendental function. The second, although less complex, requires the viscosity taken from an independent source to complete the test. It is, however, important to note that while we observe that the time evaluation of the relaxation function is well described by MCT there is no sign in the temperature dependence of any anomalies as T_c is passed. Thus, the significance of the critical temperature is still unclear.

Acknowledgements

The authors extend their thanks to K. Johansson for help in the data analysis, P. Jacobson for the sample used, and to Prof. W. Götze, Prof. Latz ande M. Fuchs, for helpful discussions regarding the theory. We also wish to thank the Swedish Natural Science Research Council for financial support.

References

1. Leutheusser E (1984) Phys Rev A29:2765
2. Bengtzelius U, Götze W, Sjölander A (1984) J Phys C19:5915
3. Sjögren L, Götze W (1992) Rep Prog Phys 55:241
4. Sidebottom DL, Bergman R, Börjesson L, Torell LM (1992) Phys Rev Lett 68:3587
5. Wang CH, Fytas G, Lilge D, Dorfmüller Th (1981) Macromol 14:1363
6. Johari GP (1986) Polymer 27:866
7. Cochrane J, Harrison G, Lamb J, Phillips DW (1980) Polymer 21:837
8. Fuchs M, Götze W, Hofacker I, Latz A (1991) J Phys Cond Matter 3:5047

Received January 10, 1992;
accepted May 25, 1992

Authors' address:

D. L. Sidebottom
Chalmers Univ. of Technology
Dept. of Physics
S-41296 Gothenburg, Sweden

Progress in Colloid & Polymer Science   Progr Colloid Polym Sci 91:46−50 (1993)

Structural relaxation characteristics of glass-forming polymeric liquids subject to transient cross-links

L. M. Torell, P. Jacobsson, D. Sidebottom, and L. Börjesson

Department of Physics, Chalmers University of Technology, Gothenburg, Sweden

Abstract: We examine the relevance of the strong−fragile classification scheme for glass-forming polymeric systems using results from Brillouin and photon correlation spectroscopy. The dynamics of salt doped polypropylene oxide (PPO) have been investigated in relation to induced local structural ordering. It is accomplished by forming complexes of $NaCF_3SO_3$–PPO in which the cations act as transient cross-links between chains. The amount of cross-links depends on salt concentration and temperature. This allows for a test of the suggested relation between structural properties and some characteristics of the relaxation function such as its non-exponential time decay and non-Arrhenius temperature dependence.

Key words: Structural relaxation − strong-fragile − light scattering

Introduction

Structural relaxation characteristics of glass-forming liquids have recently been used in the so-called strong−fragile classification scheme which relates the dynamics of the supercooled liquid state to structural properties of the glassy state [1]. The resistance of the system towards temperature-induced structural changes is taken as the basis for the classification. It suggests that as the strength of the system increases we expect less dramatic temperature behaviour above T_g of, for instance, the viscosity η or the average structural relaxation time $\langle \tau \rangle$. Among those classified as strong glass formers we find covalently bonded systems (network formers) such as the silicates. They show an almost Arrhenius temperature dependence of $\langle \tau \rangle$ and a narrow-time distribution of the relaxation decay. Using Vogel−Fulcher and Kohlraush functions to represent the temperature dependence and time decay respectively, i.e.

$$\tau = \tau_0 \exp[DT_0/(T - T_0)], \qquad (1)$$

and

$$\phi(t) = A \exp[-(t/\tau)^{\beta_K}], \quad 0 < \beta_K \leq 1, \qquad (2)$$

then strong glass formers are represented by large values of D and stretching parameters β_K (typically $D > 35$ and $\beta_K > 0.6$). At the other extreme the so-called fragile liquids demonstrate the opposite behaviour; they are highly non-Arrhenius and non-exponential, $D \approx 3$ and $\beta_K \approx 0.4$ in extreme cases. Fragile systems might locally be ordered, like the Coulomb liquids; however, there is no network structure, and they are characterized by higher degree of configurational entropy [2].

In the present work, $NaCF_3SO_3$-containing PPO systems are investigated using Brillouin scattering (BS) and photon correlation spectroscopy (PCS). In these systems topological constraints are introduced via the solvated cations which act as cross-links between oxygens of adjacent chains [3–12]. PPO is a Lewis base; the ether oxygens are donors and the salt cations acceptors; the cations are solvated. The anion is only weakly solvated (if at all) and is delocalized among the polymer chains. The cation is generally co-ordinated to several polyethers and can thereby establish a cross-link between the adjacent chains. The so solvated cations introduce compressional effects (electrostriction) and the densification increases, the lower the polarizability and the higher the charge density of the cation. The density of cross-links can be controlled from the salt concentration and/or via temperature. Cationic cross-links in salt–polymer complexes, and the effect of such cross-links on density, viscosity, and glass transition temperature, have been reported by many authors [3–10]. A model has recently been developed which discusses the compressional effects and the temperature-dependent ordering imposed on the system via the solvated cations [11–12]. By introducing temperature-dependent cationic ordering we will investigate whether in polymers, like in low molecular weight glass formers, it is the resistance towards structural changes which determines their strong–fragile classification.

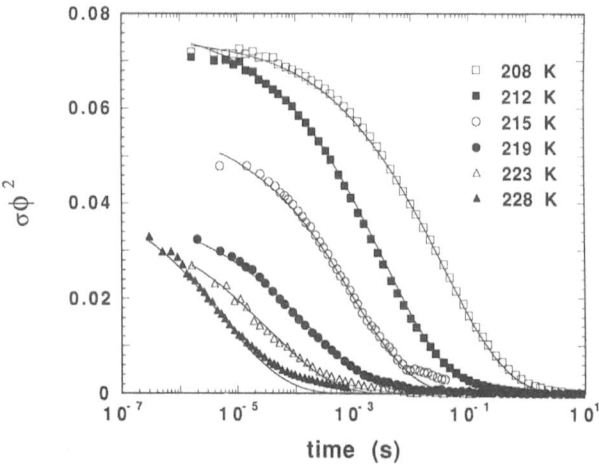

Fig. 1. PCS autocorrelation function $\sigma\phi^2(t)$ of PPG 4000 from 208 to 228 K. Solid lines represent KWW fits using Eq. (2)

Experimental

Details of the preparation of the salt–polymer complexes PPO–NaCF$_3$SO$_3$ are given elsewhere [10, 13]. The photon correlator (ALV 3000) employs a pseudo-logarithmic time-base to achieve an eight decade wide measure of the autocorrelation function in a single run. Details of the PCS set up are given in ref. [14]. The Brillouin equipment includes a triple passed piezoelectrically scanned Fabry–Perot interferometer (Burleigh RC-110) stabilized by a feedback system (DAS 10) and is described elsewhere [15]. Argon ion lasers (Spectra Physics) were used as light (488 nm) sources.

Results and discussion

The α-process of PPO

First we will determine the relaxation characteristics of uncomplexed PPO and in Fig. 1 we display some PCS recordings. The shape of the relaxation decay has been investigated in detail over a broader temperature range using mode coupling theory [16] and results are given in ref. [14]. The analysis reveals that two types of relaxation processes occur above T_g. Here, we will only address the main relaxation, the α-process related to the glass transition, and it is the dominating process in Fig. 1. The other process is identified as the α' and is only noticeable at higher temperatures [14].

To determine the temperature dependence of the relaxation time and the stretching parameter of the α-process, we fitted Eq. (2) to the measured correlation functions; see solid curves in Fig. 1. A stretching parameter of $\beta_K = 0.39 \pm 0.03$ was found independent of temperature in the range 208–228 K. The obtained relaxation times were converted into average relaxation times according to $\langle\tau\rangle = (\tau/\beta_K)\Gamma(1/\beta_K)$, where $\Gamma(x)$ is the gamma function. Results of $\langle\tau\rangle$ are displayed in Fig. 2. The relaxation time obtained at a much higher probe frequency by Brillouin scattering [8] is also included in Fig. 2. In the Brillouin study it was shown that the relaxation time was independent of molecular weight for molecular weights in the range 400–10 000. This was also found in the photon correlation study by Wang et al. [17]. A single Vogel–Fulcher law with $D \approx 7$ in Eq. (1) can represent all the present and previous [8, 17] data over the whole temperature and molecular weight range and it is presented as a solid line in Fig. 2. Moreover, the same value for the stretching parameter as obtained in the present study was reported for PPO at Brillouin frequencies and it was shown to be independent of molecular weight [8].

The α-process of cationic cross-linked PPO–NaCF$_3$SO$_3$

Next we investigate the effects of transient ionic cross-links on the structural relaxation dynamics using our Brillouin scattering data. The spectra reveal strong relaxation effects and the results are discussed in detail in ref. [13]. Here we only focus on the structural relaxation time τ and the stretching parameter β_K of Eq. (2).

An average value of τ can be obtained from the peak condition $\omega\tau \sim 1$, where $\omega = \pi\nu_B$ and ν_B is

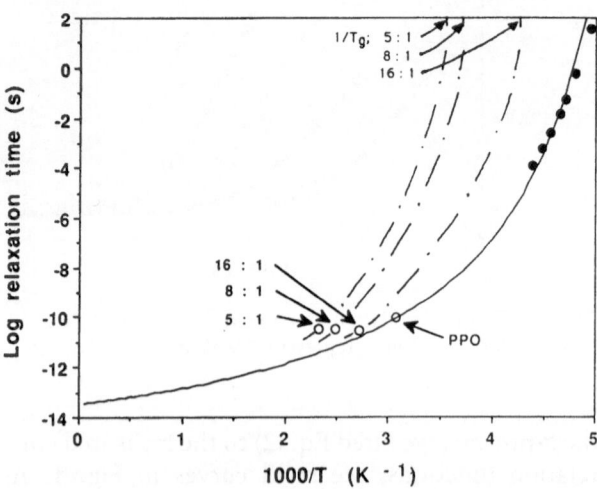

Fig. 2. Structural relaxation time vs. inverse temperature for uncomplexed PPO and for PPO–NaCF$_3$SO$_3$ complexes of concentrations O : M 16 : 1, 8 : 1, and 5 : 1. Open symbols show results obtained from Brillouin scattering, closed symbols are from photon correlation measurements. The solid line represents a best fit to present data of PPO (molecular weight 4000) and to previous data obtained for PPO in the molecular weight range 400–10 000 (see refs. [8] and [17]). O : M is the monomer/salt molar ratio. T_g is indicated for the various salt–polymer complexes (from ref. [9])

the Brillouin frequency shift at the temperature of the absorption peak, i.e., the temperature at which the Brillouin lines demonstrate maximum broadening. In Fig. 2 we give results for the complexes and it can immediately be seen that the structural relaxation process in the Brillouin range is very sensitive to the amount of dopant salt. For the almost constant Brillouin frequency ($\sim$ 5 GHz) the measured average relaxation times (also almost constant, $\sim 5 \cdot 10^{-11}$ s) are shifted to higher temperatures with increasing salt concentration (Fig. 2). The change of the relaxation range is also indicated by increased values of T_g [9] (see Fig. 2); T_g is generally taken as the temperature at which the relaxation time is $\sim$ 100 s [1]. This implies that the whole relaxation pattern is shifted towards longer times and it is schematically represented in Fig. 2 by dashed-dotted lines. The slowing down of the intrachain motions can be explained by that cationic cross-links are established, see inset of Fig. 3. This is also demonstrated by increased values of the hypersonic velocity [13] and by the behavior of the Raman D-LAM (disordered longitudinal acoustic mode) [18]. In this context it is interesting to note that in uncomplexed PPO the time scale for local structural rearrangements probed by light scattering is unchanged upon polymerization (see last paragraph above), despite the fact that on the macroscopic scale the viscosity increases linearly with the molecular weight up to the chain entangle-

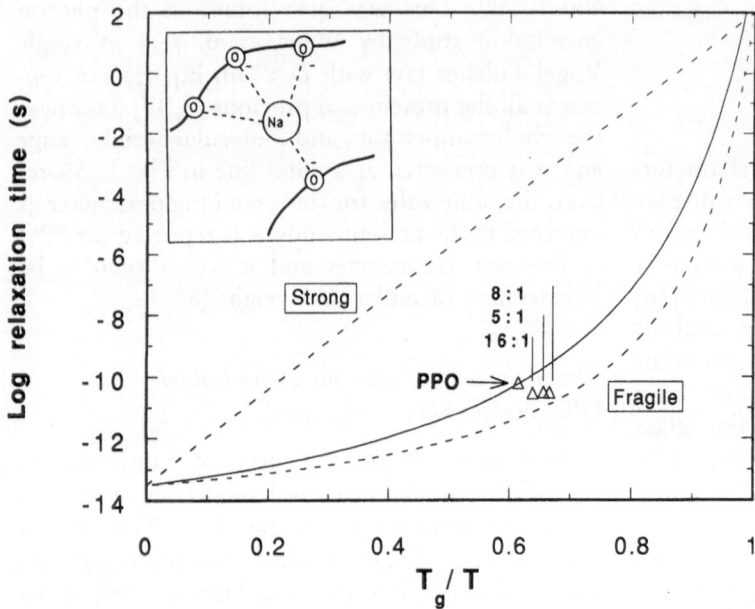

Fig. 3. Structural relaxation time vs. scaled temperatures T_g/T for uncomplexed PPO and for PPO–NaCF$_3$SO$_3$ (M_w 4000) complexes of concentrations O : M = 16 : 1, 8 : 1, and 5 : 1. Solid line is from Fig. 2 and represents data for uncomplexed PPO. Dashed lines represent "strong" and "fragile" extremes, respectively. Inset shows a schematic picture of a transient cross-link between two polymer chains via a sodium ion

ment point. The macroscopic viscosity also increases when salt is solvated by the polymer [19]; however, in this case Fig. 2 shows that now the local structural relaxation time is affected.

We now turn to the fragile–strong representation by plotting the data on Fig. 2 in the T_g reduced plot of Fig. 3. Such a scaling reduces the differences between the various salt-containing complexes. However, we note that they all depart from the dynamics of the uncomplexed system and move towards the fragile end. The effect can be explained by the temperature dependence of the cross-links as discussed in the following. Raman investigations of the same systems show that close to the glass transition about 85% of the ions are solvated, whereas only 30% of the ions are dissolved at about 100 K above T_g [20]. Since it is the solvated cations which act as cross-linking centres, the Raman data imply that as temperature increases the number of cationic cross-links rapidly decreases. The weakened cation–polymer interaction in turn results in increased cation–anion interactions [20], the anions being only weakly co-ordinated to the polymer chain. Such neutral cation-anion pairs may act as plasticiser of the system and increase the local flexibility. Thus, while the cations introduce crosslinks in the low temperature range and thereby stiffen the system, the same cations may form plasticising pairs and increase the segmental mobility in the high temperature range. This implies that adding salt to PPO introduces a cross-linked network which rapidly breaks down and accordingly the relaxation function moves towards the fragile end. Such a conclusion is further supported by the concentration dependence, noting that in Fig. 3 the system of salt concentration 8:1 shows maximum departure from the behaviour of the uncomplexed PPO. The corresponding D parameter of Eq. (1) is for this system ≈ 5 which is considerably smaller than that of the uncomplexed PPO, ≈ 7, and not far away from the fragile extreme value ≈ 3. In the 8:1 concentration range the amount of solvated ions is reported to be at maximum [18, 20]. Thus, we expect maximum density of cross-links in this system at lower temperatures and, therefore, the largest departure in the T_g reduced plot from the curve representing uncomplexed PPO as is indeed observed in Fig. 2.

Next we focus on the stretching parameter β_K which can be determined from a fit of the acoustic absorption coefficient α. The absorption is

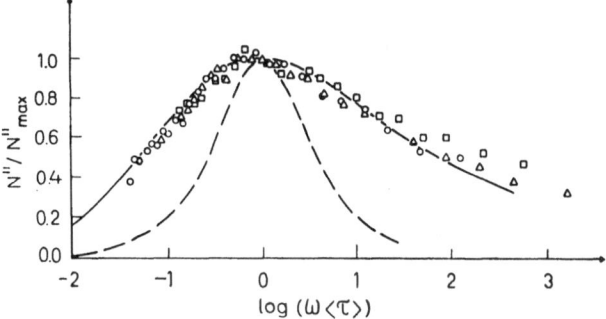

Fig. 4. Imaginary part, N'', of the reduced elastic modulus vs. $\log[\omega\langle\tau\rangle]$ for PPG–NaCF$_3$SO$_3$ complexes of concentrations O:M = 16:1, (○), 8:1 (△), and 5:1 (□). Solid line represents fits of Kohlrausch–Williams–Watts function with $\beta_K = 0.35$. Dashed curve represents Debye relaxation

related to the Brillouin half-width Γ_B through $\alpha = \pi\Gamma_B/v$ where v is the sound velocity obtained from the Brillouin shift. The results for the absorption coefficient are given in the reduced modulus plot of N'' in Fig. 4, where N'' has been calculated according to

$$N''(\omega) = [2\alpha(\omega)v^3(\omega)]/[\omega(v_\infty^2 - v_0^2)] . \qquad (3)$$

Here v_0 and v_∞ are the limiting low and high-frequency velocity values, respectively. Analytically, N'' is expressed by the relaxation function $\phi(\tau)$ according to

$$N''(\omega\tau) = \omega\tau \int_0^\infty \cos(\omega\tau)\,\phi(t/\tau)\,d(t/\tau) . \qquad (4)$$

Assuming a KWW relaxation function, we obtain a stretching parameter $\beta_K \approx 0.35$. We note that this is indeed smaller than the value $\beta_K \approx 0.40$ of uncomplexed PPO. Within the accuracy of the data no differences were observed between the different complexes. The overall smaller β_K value of the salt-containing systems supports the findings of Fig. 3 that the relaxation dynamics is shifted towards the fragile end in favour of the proposed relation between the non-exponentiality of the relaxation function and the departure from Arrhenius behaviour.

In summary, by introducing temperature-dependent network connectivity via salt complexation of PPO we find that the system follows the general pattern of the strong–fragile classification scheme. Moreover, it is found that increased fragility is accompanied by increased stretching of the relaxation time decay as has been suggested.

Acknowledgements

This work was carried out with support from the Swedish Natural Science Research Council.

References

1. Angell CA (1985) In: Ngai K, Wright GB (eds) Relaxation in Compl. Systems. National Technical Information Service, U.S. Department of Commerce, Springfield, VA22161, 1
2. Angell CA (1988) Phys Chem Solids 49:8634
3. Wetton RE, James DB, Whithing W (1976) Polym Sci Polym Lett Ed 14:577; (1979) Polymer 20:187
4. Berthier C, Gorechi W, Minier M, Armand M, Chakagno J, Rigaud P (1983) Solid State Ionics 11:19
5. Papke B, Ratner M, Scriver D (1981) J Phys Chem Solids 42:493
6. Ratner M (1988) In: MacCallum JR, Vincent CA (eds) Polymer Electrolyte Review I, Ch. 7. Elsevier, New York
7. McLin MG, Angell CA (1991) J Phys Chem 95:9464
8. Börjesson L, Torell LM, Stevens JR (1987) Polymer 28:1803; (1987) Physica Scripta 35:692
9. Stevens JR, Schantz S (1988) Polymer Comm. 29:330; Wixwat W, Fu Y, Stevens JR (1991) Polymer 32:1181
10. Torell LM, Jacobsson P, Sidebottom D, Petersen G, Solid State Ionics, in press
11. Ratner M, Nitzan A (1989) Faraday Discuss Chem Soc 88:19
12. Olender R, Nitzan A (1992) Electrochimica Acta 37:1505
13. Sandahl J, Schantz S, Börjesson L, Torell LM, Stevens JR (1989) J Chem Phys 81:2
14. Sidebottom D, Bergman R, Börjesson L, Torell LM (1992) Phys Rev Lett 68:3587
15. Börjesson L (1987) Phys Rev B 36:4600
16. Fuchs M, Götze W, Hofacker I, Latz A (1991) J Phys: Condensed Matter 3:5047
17. Wang CH, Fytas G, Lilge D, Dorfmüller Th (1981) Macromolecules 14:1363
18. Schantz S, Sandahl J, Börjesson L, Torell LM, Stevens JR (1988) Solid State Ionics 28–30:1047
19. Torell LM, Angell CA (1988) British Polymer J 20:173
20. Kakihana M, Schantz S, Torell LM (1990) J Chem Phys 92:6271

Received January 26, 1992;
accepted June 25, 1992

Authors' address:

L. M. Torell
Chalmers University of Technology
Dept. of Physics
S-412 96 Gothenburg, Sweden

Progress in Colloid & Polymer Science   Progr Colloid Polym Sci 91:51–54 (1993)

Synchrotron SAXS studies of segmented polyurethanes

B. Chu and Y. Li

Department of Chemistry, State University of New York at Stony Brook, Long Island, USA

Abstract: It has been reported by us, for the first time, that segmented polyurethanes (SPUs) based on 4,4'-diphenylmethane diisocyanate (MDI) and 1,4-butanediol (BD) as the hard segment could form spherulite structures upon melt-quenching to an annealing temperature (T_A) above 120 °C. Synchrotron small-angle x-ray scattering (SAXS) experiments were performed to investigate the microphase separated structures corresponding to the spherulite formation. With increasing hard-segment content, both the integrated scattered intensity (Q) and the interdomain spacing (d) increased first and then decreased when the hard segment was above 50%, possibly due to phase inversion. With increasing T_A, the Q values increased first, reached a maximum at $\sim$ 107 °C, and then decreased with further increase in T_A; the d values showed a monotonic increase. The kinetics of phase separation was investigated after the samples were quenched from the melt to 140 °C. Interdomain spacing remained unchanged throughout the whole process. The results indicated that the hard-segment mobility and the system viscosity were the two key factors controlling the phase structure.

Key words: Synchrotron SAXS – segmented polyurethanes – microphase separation – annealing effects – phase separation kinetics

Introduction

One of the unique characteristics of thermoplastic segmented polyurethane (SPU) elastomers is that the crosslinking in the system is physical in nature. Unlike chemically crosslinked conventional rubbers, SPUs can be processed by melting them at high temperatures. Therefore, it is possible to develop materials with entirely different structures and properties by utilizing different thermal treatment. However, the thermal effects on the structure–property relationship of SPU starting from the melt state have not been investigated carefully using a structural scale from nanometers up to microns. The spherulite structure could be observed from SPUs consisting of MDI and BD as the hard segment after the melts were quenched to annealing temperatures (T_A) above 120 °C [1]. The high flux of a synchrotron x-ray source enabled us to follow the phase separation kinetics after the melts were quenched to T_A. The effects of thermal treatment on the structure–property relationship of SPUs are reported.

Experimental

The SPU samples were synthesized by a two-step bulk polymerization. The hard segment consists of 4,4'-diphenylmethane diisocyanate (MDI) and 1,4-butanediol (BD). The soft segment is polytetramethylene oxide end-capped with polypropylene (PPO-PTMO) ($M_n = 1000$, not crystallizable in this study). The samples are denoted as PPO-PTMO-PU-*xx* where *xx* represents the percentage of the MDI–BD hard-segment content. Annealing experiments were performed in vacuum.

The synchrotron SAXS experiments were performed at SUNY X3A2 beamline, National Synchrotron Light Source (NSLS), Brookhaven National Laboratory (BNL), by using a modified Kratky collimation system along with a Braun linear position-sensitive detector, and at the SAXS facility, SUNY, Stony Brook. The wavelength of x-rays used was 0.154 nm. The sample-to-detector distance was 1330 mm. The beam size at the sample position was $\sim$ 0.2 × 2 mm². The synchrotron beam was *focused* onto the beam stop which was ~ 40 mm in front of the detector having a receiving window width of ~ 2 mm. Therefore, smearing effects on the SAXS profiles were negligible with synchrotron x-rays. All data presented here were from the SAXS instrument at NSLS.

Results and discussion

In a SAXS experiment, the integrated scattered intensity, Q (also called invariant), is related to the mean square electron density fluctuations and can be expressed as

$$Q = \int q^2 I(q)\, dq , \tag{1}$$

where q is the magnitude of the scattering vector. For an ideal two-phase system with a sharp interphase

$$Q_i = c\phi_h \phi_s (\rho_h - \rho_s)^2 , \tag{2}$$

where c is a constant and ϕ_i and ρ_i are the volume fraction and the electron density of the ith phase, respectively, with h and s denoting hard and soft segment. Therefore, Q/Q_i is an approximate indication of the degree of phase separation. We noted that Eq. (2) is valid only for an ideal two-phase system, while our SPU systems might involve diffusive interphase boundaries. Our set-up at NSLS might not have reached high enough q values

for an accurate evaluation of the effects due to the diffusive interphase boundaries on the Q/Q_i values. However, preliminary analysis showed that our conclusions in the present work remained unchanged by including these effects. The error introduced could be reduced further, since our discussion was based on the *ratio* of integrated scattered intensity (Q/Q_i). Furthermore, we are only interested in the *relative changes* due to the thermal treatment.

The hard-segment domain in the SPUs is believed to have a lamellar structure, at least on a localized scale. The interdomain spacing d can be obtained by using the Bragg equation from Lorentz-corrected scattering curves (Iq^2 vs. q)

$$d = 2\pi/q_{max} , \tag{3}$$

where q_{max} denotes the location of the scattering peak. A more reasonable way is to use the correlation function analysis where the first non-zero maximum is a good estimate of d. Both methods gave the same conclusion in this study. The results from Eq. (3) and the one-dimensional correlation function analysis agree with each other to within $\sim 5\%$. Therefore, Eq. (3) was used to obtain the interdomain spacing in the present study.

With increasing hard-segment content, both Q/Q_i and d increased first and then decreased, as shown in Fig. 1, when the hard-segment content is

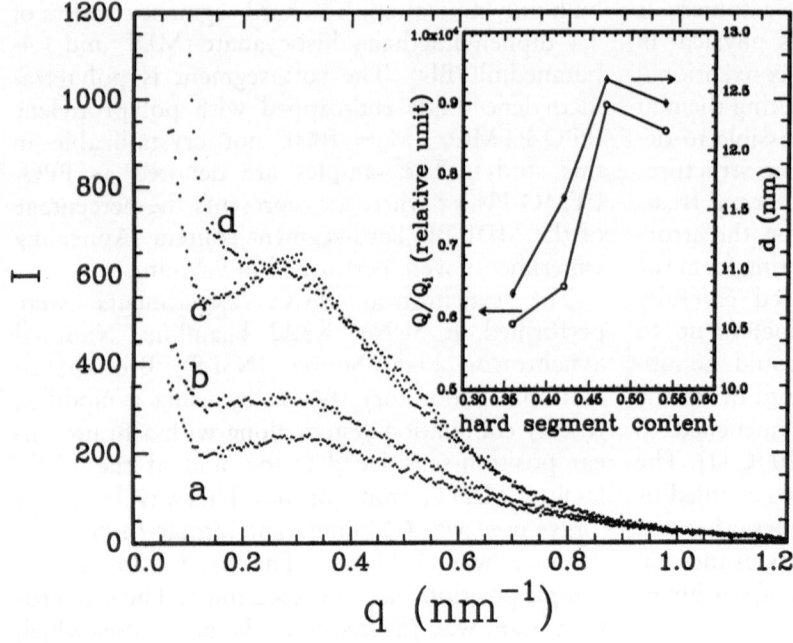

Fig. 1. SAXS profiles of PPO-PTMO-PU series quenched from melts to 80°C and annealed for 48 h: (a) PPO-PTMO-PU-36; (b) PPO-PTMO-PU-42; (c) PPO-PTMO-PU-47; (d) PPO-PTMO-PU-54. The dependence of Q/Q_i and d upon hard-segment content is shown in the inset. Measurements were performed at room temperatures

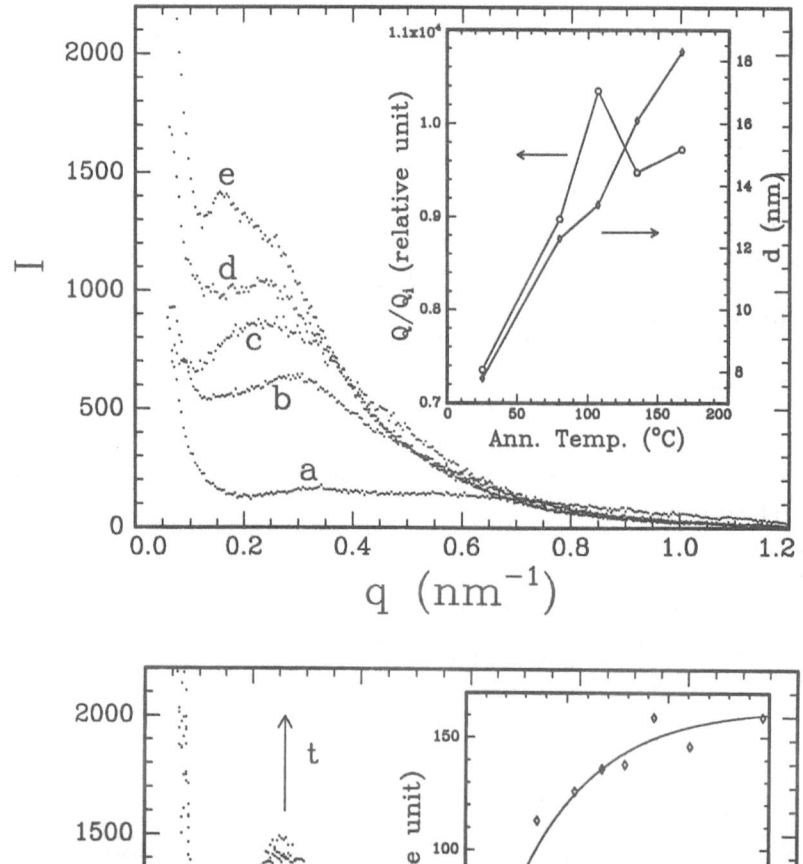

Fig. 2. SAXS profiles of PPO-PTMO-PU-47 quenched from melts to (a) 25 °C; (b) 80 °C; (c) 107 °C; (d) 135 °C; (e) 167 °C. The dependence of Q/Q_i and d upon T_A is shown in the inset. Measurements were performed at room temperatures. The difference in SAXS profiles measured at room temperatures and at annealing temperatures could be attributed as due to the thermal expansion of the two phases. Secondary phase separation was not significant

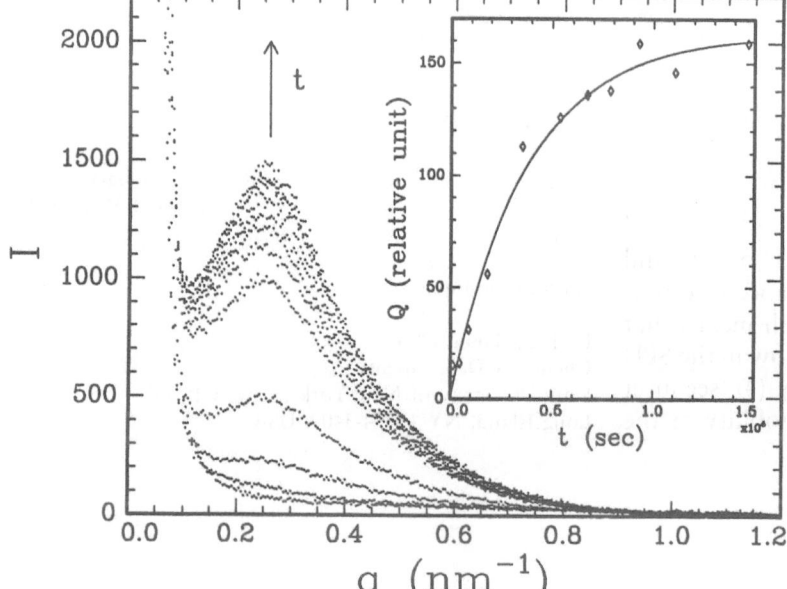

Fig. 3. SAXS profiles of PPO-PTMO-PU-50 as a function of elapsed time after the melt (230 °C) was quenched to 140 °C. A fitting of $Q \sim t$ to Eq. (4) is shown in the inset. Measurements were performed at 140 °C

above 50%, possibly due to phase inversion. The increase in d could be due to the increase in the hard-segment length with increasing hard-segment content. The incompatibility between hard and soft segment increased with increasing hard-segment length [2], which could explain the increase in Q/Q_i.

Figure 2 shows that as T_A was increased, d showed a monotonic increase. Q/Q_i increased first, reached a maximum at ~ 107 °C, and then decreased with further increase in T_A.

In SPU systems, the mobility of the hard-segment chains is a key factor. At low T_A, the low hard-segment mobility and the high system viscosity prevented good phase separation and crystallization. At high T_A, the increase in mobility and the decrease in viscosity resulted in an increase in Q/Q_i. A temperature of 107 °C was slightly above

T_g of the hard segment domain. The increasing mobility promoted a maximum in Q/Q_i. With a further increase in T_A, the short hard segment could be dissolved in the soft segment matrix (phase mixing), thus reducing the phase contrast. The monotonic increase in d could be explained by extending the Koberstein–Stein coiled/folded chain model [3], i.e., the hard-segment chains are basically coiled or folded. The hard-segment chains extended gradually with increasing T_A. After the hard-segment chains had extended to a certain degree, crystallization between the chains became possible. At T_A above 120 °C, the crystallites could be arranged into spherulites.

The phase separation kinetics of PPO-PTMO-PU-50 was followed by synchrotron SAXS after the melt was quenched to 140 °C (Fig. 3). The d values remained unchanged throughout this process, indicating that the degree of coiling or folding was controlled only by T_A. The scattered intensity showed monotonic increase with elapsed time at all SAXS q values of our experiments. The Q/Q_i values as a function of elapsed time can be fitted by an equation of relaxation

$$[(Q/Q_i)_\infty - (Q/Q_i)_t]/[(Q/Q_i)_\infty - (Q/Q_i)_0]$$
$$= (Q_\infty - Q_t)/(Q_\infty - Q_0)$$
$$= \exp(-t/\tau) , \tag{4}$$

where τ is the relaxation time and ∞, t, and 0 denotes elapsed time at infinity, t, and 0, respectively. The τ value is $3.7 \cdot 10^3$ s, which means that the phase separation process is very slow in the SPU system. A good fitting by means of Eq. (4) (see inset in Fig. 3) once again suggests that mobility of the hard segment and viscosity of the system are key factors controlling the structural changes in the SPU system.

Concluding remarks

The annealing temperature has a significant effect on the phase structure. The phase separation process is very slow (accomplished in hours), making the thermal history a very important factor in SPU processing procedures.

Acknowledgements

We thank Prof. J. Liu and Mr. T. Gao for their assistance in this work. B.C. acknowledges the financial support of this project by the U.S. Department of Energy (DEFG0286ER45237A004 and DEFG0589ER75515).

References

1. Li Y, Liu J, Yang H, Ma D, Chu B (1991) Polym Mater Sci Eng 65:297–298
2. Krause S (1973) In: Burke J, Weiss V (eds) Block and Graft Copolymers. Syracuse University Press, New York, pp 143
3. Koberstein J, Stein R (1983) J Polym Sci Polym Phys Ed 21:1439–1472

Received January 23, 1992;
accepted May 26, 1992

Authors' address:

Prof. Benjamin Chu
Chemistry Department
State University of New York at Stony Brook
Long Island, NY 11794-3400, USA

Investigations of self- and tracer diffusion in poly(ethylene oxides) and in blends of poly(dimethyl/ethylmethyl siloxanes) with the pulsed field gradient NMR

G. Fleischer

Fachbereich Physik, WB Polymerphysik, Universität Leipzig, FRG

Abstract: The pulsed field gradient NMR was used to measure the self-diffusion coefficients of poly(ethylene oxides) in solution and melt and the tracer-diffusion coefficients of both components in the blend of poly(ethylmethylsiloxane)/poly(dimethylsiloxane). In the melts of lower molecular weight poly(ethylene oxide) aggregation behaviour as reported in literature (VA Sevreugin et al. (1986) Polym 27:290) within the time scale of the experiment could be excluded. For high molecular weight solutions diffusion within the tube was detected and $\langle z^2 \rangle \sim t^\kappa$ with $\kappa = 0.3 \ldots 0.4$ was found. In polysiloxane blends the tracer diffusion has no discontinuity at the temperature of spinodal decomposition. The tracer-diffusion coefficients are to some extent averaged in the blend in comparison with the self-diffusion coefficients in the pure polymers.

Key words: Self-diffusion – tracer diffusion – PFG–NMR – poly(ethylene oxide) – poly(siloxanes)

The quantity measured in pulsed field gradient NMR (PFG–NMR), the spin echo damping Ψ, is the Fourier transform of the self-correlation function $P(z, t)$ (in one dimension z) of the spin-containing segments of the polymer chain with respect to the "wave vector" $\gamma \delta g$:

$$\Psi(\gamma \delta g) = \int P(z, t) \exp(i\gamma \delta g z) \, dz \, . \qquad (1)$$

γ is the gyromagnetic ratio, δ is the width and g is the magnitude of the two-field gradient pulses separated by the time $t = \Delta$. With a Gaussian self-correlation function, in the diffusion limit, one arrives at

$$\Psi = \exp(-\gamma^2 \delta^2 g^2 \langle z^2 \rangle / 2) \, , \qquad (2)$$

the well-known Tanner–Stejskal equation. The self-diffusion coefficient D is $D = \langle z^2 \rangle / 2\Delta$ [1]. If the inverse of the "wave vector" is in the order of a characteristic length of the system one can measure non-Fickian diffusion, e.g. restricted diffusion. For polymer molecules at distances smaller than the Flory radius R_F the mean square displacements $\langle z^2 \rangle$ of the spin-containing segments

are proportional to the diffusion time Δ with a power smaller than one, for curvilinear diffusion of the whole chain within the tube, e.g. $\langle z^2 \rangle \sim \Delta^{0.5}$ [2]. In this case the self-correlation function of the segments remains Gaussian and Eq. (2) applies. The pulsed field gradient NMR is quite analogous to the incoherent inelastic neutron scattering but has larger time and space scales.

We have used very large field gradient pulses and have $(\gamma \delta g)^{-1}$ extended to values as small as about 300 Å. This was possible by using the large field gradients existing in the stray field of commercial cryomagnets [3].

We have measured the self-diffusion in melts and in concentrated solutions of a high molecular weight poly(ethylene oxide) and in blends of poly(dimethyl/ethylmethyl siloxanes) at the critical concentration. In the entangled PEO solutions (PEO $5 \cdot 10^6$ 5% in D_2O and 50% in C_6D_6), we have detected the cross-over from free diffusion of the segments for distances larger than the Flory radius to restricted segmental motions (within the tube) for distances smaller than the Flory radius of

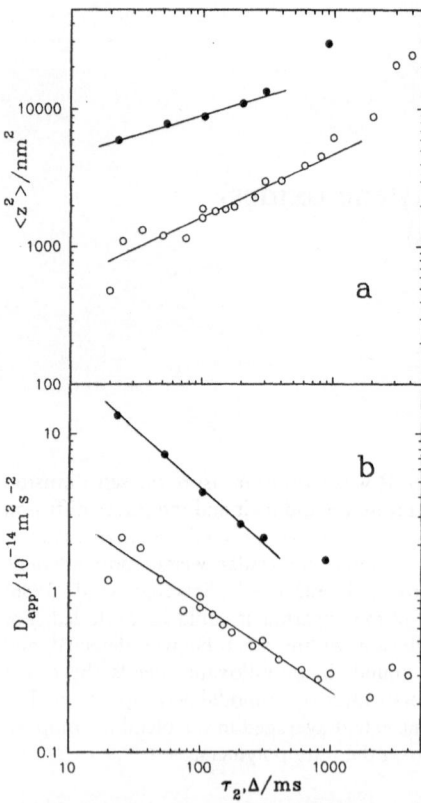

Fig. 1. Mean square displacement $\langle z^2 \rangle$ and apparent diffusion coefficient D_{app} vs. the diffusion time $\Delta = \tau_2$ for poly(ethylene oxide) $5 \cdot 10^6$. $\bigcirc$: dissolved in C_6D_6 (50%) and measured at $T = 61\,°C$ with the NMR in the stray field of a cryomagnet, $\bullet$: dissolved in D_2O (5%) and measured at $T = 25\,°C$ with the (conventional) PFG–NMR

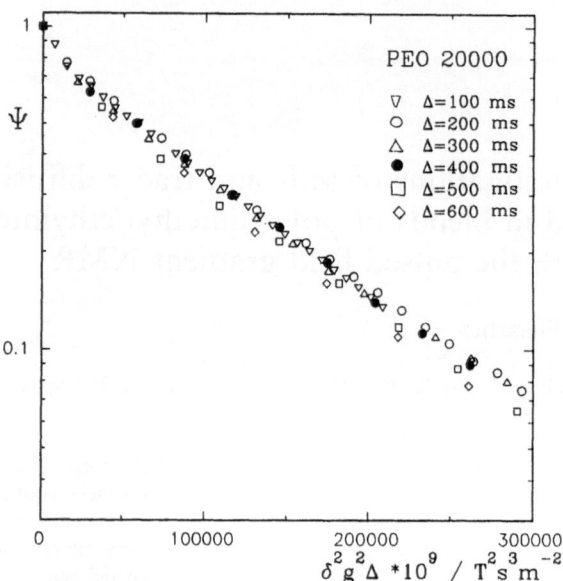

Fig. 2. Echo attenuation for the PEO 20 000 at $T = 100\,°C$ for different diffusion times Δ

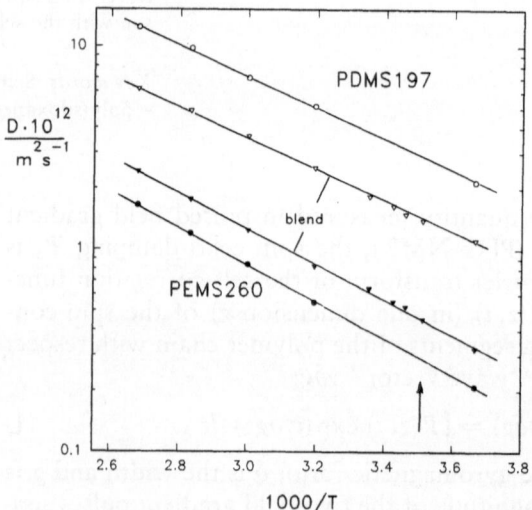

Fig. 3. Arrhenius plot for the self-diffusion coefficients of PEMS 260 and PDMS 197 in the pure polymers and the tracer-diffusion coefficients in the blend at the critical concentration of $\Phi_{PEMS} = 0.445$. The critical temperature is indicated by an arrow

the coils where $\langle z^2 \rangle \sim \Delta^{0.5}$ approximately holds. The result is shown in Fig. 1. The cross-over occurs at the root-mean-square displacement $\langle z^2 \rangle^{0.5}$ of about 100 nm which is equal to the Flory radius of the polymer.

In the melt of poly(ethylene oxides) with molecular weights between 6000 and 40 000 cluster formation within the time scale of the experiment (a few ms up to 2 s) was reported from Sevreugin et al. [4]. This could not be confirmed in our investigations. Careful measurements of the echo attenuations for poly(ethylene oxides) with the molecular weights 6000, 20 000 and 100 000 show no dependence of Ψ on the diffusion time Δ and, hence, no dynamic cluster formation. Therefore, cluster formation in this polymer melt must be excluded. An example of the measured echo attenuations is shown in Fig. 2.

In the polysiloxane mixtures the tracer diffusion coefficients of each of the two components of the system have been measured in dependence on the temperature beginning at $T \gg T_{cr}$ down to temperatures below the critical temperature T_{cr} where spinodal decomposition takes place. The result for the blend PEMS 260/PDMS 197 (the numbers denote the polymerization index) is shown in Fig. 3. We observe an averaging of the monomeric friction coefficients in the mixture in comparison to the pure polymers. In the lower molecular weight system PDMS 80/PEMS 90 the monomeric friction coefficients are found to be equal within experimental accuracy, and with this result the "fast-mode" and "slow-mode" ansatz for interdiffusion become indistinguishable [5]. At spinodal decomposition, where the interdiffusion shows the critical slowing down to zero diffusion, the tracer diffusion coefficients have no discontinuity.

References

1. Kärger J, Pfeifer H, Heink W (1988) Adv Magn Res 12:1–89
2. de Gennes PG (1978) Scaling Concepts in Polymer Physics. Cornell Univ Press, Ithaca, NY
3. Fleischer G, Fujara F (1992) Macromolecules 25:4210–4212
4. Sevreugin VA, Skirda VD, Maklakov AI (1986) Polym 27:290–295
5. Momper B (1990) Dissertation, Universität Mainz

Received November 4, 1991;
accepted May 21, 1992

Authors' address:

Dr. Gerald Fleischer
Universität Leipzig
Fachbereich Physik
WB Polymerphysik
Linnéstraße 5
D-O-7010 Leipzig, FRG

Progress in Colloid & Polymer Science   Progr Colloid Polym Sci 91:58–60 (1993)

Segmental relaxation in a symmetric poly(styrene-b-methylphenylsiloxane) copolymer in the disordered phase

B. Gerharz[1]), S. Vogt[1]), E. W. Fischer[1]), and G. Fytas[2])

[1]) Max-Planck-Institut für Polymerforschung, Mainz, FRG
[2]) Foundation for Research and Technology Hellas, Heraklion, Crete, Greece

Abstract: Orientational dynamics in a homogeneous diblock copolymer of polystyrene (PS) and poly(methylphenylsiloxane) (PMPS) with PS composition $f_{PS} = 0.58$ and a single but very broad glass transition temperature $T_g = 303 \pm 45$ K have been studied by photon correlation (PCS) and dielectric (DS) spectroscopy over the broad temperature range from 273 to 373 K. Two distinct primary relaxation processes associated with the PS-like and PMPS-like segmental dynamics can be resolved by PCS and DS techniques. The corresponding distributions of relaxation times exhibit a broad shape with temperature-dependent breadths.

Key words: Primary relaxation – composition fluctuations – depolarized Rayleigh scattering – dielectric spectroscopy – diblock copolymer

Introduction

The phase behavior of a diblock copolymer melt is determined by the dimensionless product χN of the segment–segment interaction parameter χ and the total degree of polymerization N, as well as by the composition f [1]. In the mean field theory, symmetric block copolymers are microphase separated when $(\chi N) > 10.495$. The fluctuations in the average composition f are predicted to be very small and a microphase separated sample should exhibit two distinct glass transitions in contrast to the single, usually broad, glass transition of a disordered sample. Inclusion, however, of fluctuation effects [2] in the mean field theory increases the amplitude of the local composition fluctuations $\langle |\delta f|^2 \rangle$. Segmental dynamics should be seriously affected by such structural heterogeneities and, hence, employment of experimental techniques probing local motions would be very helpful. This short paper reports on segmental reorientation in a homogeneous diblock copolymer utilizing photon correlation spectroscopy (PCS) in the depolarized geometry and dielectric spectroscopy (DS).

Experimental

The diblock polystyrene–poly(methylphenylsiloxane) (PS–PMPS) with a PS volume fraction standardized by PS $f_{PS} = 0.58$, $M_w = 11\,760$ g/mol and $M_w/M_n = 1.12$ displays a broad single $T_g = 303$ K and breadth $\Delta T_g = 90$ K. Over the temperature (T) range from 273 to 373 K χN was estimated to vary between $\chi N = 3.18$ and $\chi N = 2.85$ [3]. A dust-free sample of high optical quality was obtained by filtering under pressure through a 0.45 μm Durapore filter (Millipore). The time correlation function $G(t)$ of the depolarized light-scattering intensity was measured with a full correlator (AVL-5000) in the time range 10^{-6}–10^3 s at different temperatures between 298 and 353 K. The desired orientational relaxation function $C(t)$ was computed from $|C(t)|^2 = (G(t) - 1)/b$, where b is the instrumental factor [4]. Figure 1 (upper part) shows $|C(t)|^2$ as obtained for PS–PMPS at 308 K. It is worth mentioning the broad shape and the value of the plateau of the $C(t)$ at short times. The optical anisotropy $\langle \gamma^2 \rangle / x$ per monomer amounts to 38 and 28 Å^6, respectively, for PS[5] and PMPS [6] at 298 K. Therefore, it seems that

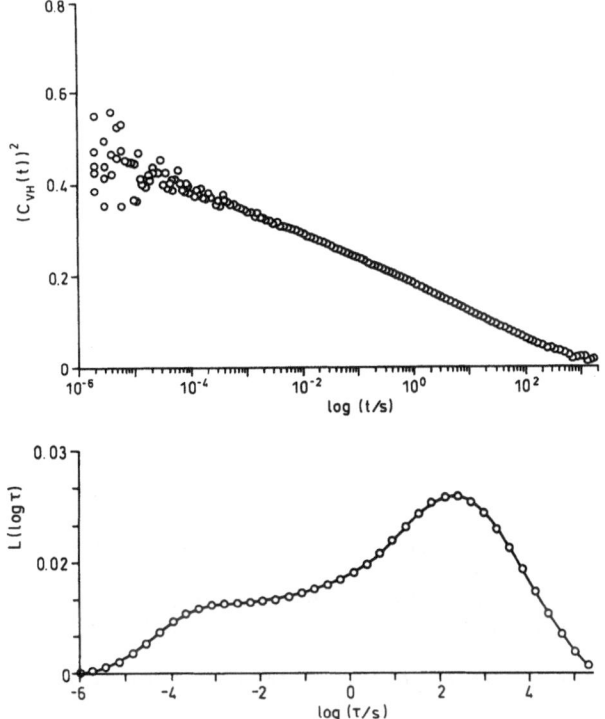

Fig. 1. Orientational time correlation function $|C(t)|^2$ for the PS–PMPS diblock copolymer at 308 K (upper part) and the corresponding distribution of relaxation times obtained from the inversion of the $C(t)$ (lower part)

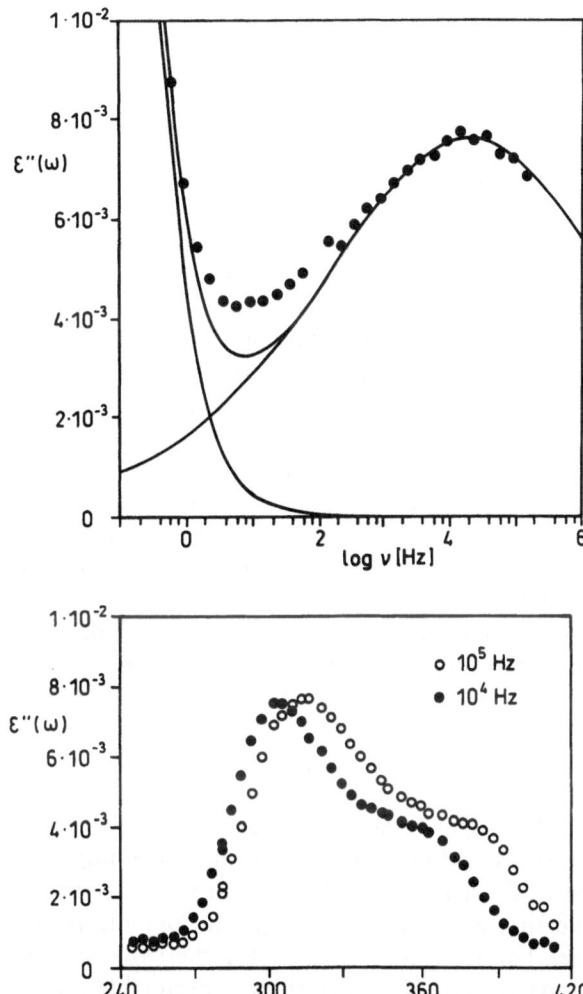

Fig. 2. Dielectric loss ε'' versus frequency for the PS–PMPS sample at 305 K. The solid line denotes the sum of a single HN fit and a conductivity fit to the experimental $\varepsilon''(\omega)$ which fails to describe the experimental data (upper part) and temperature dependence of $\varepsilon''(\omega)$ for the PS–PMPS copolymer (lower part) at two different frequencies ($\bullet$: 10^4 Hz, $\circ$: 10^5 Hz)

$C(t)$ in Fig. 1 is due mainly to PS segmental dynamics. The dielectric permittivity $\varepsilon^*(\omega)$ was measured using a frequency response analyzer [7] (Solarton–Schlumberger FRA 1260) in the frequency range 10^{-1}–10^6 Hz and over the temperature range 243–413 K. The relaxation strength $\Delta\varepsilon$ of the two constituent homopolymers is very different. The ratio $\Delta\varepsilon$ (PMPS)/$\Delta\varepsilon$(PS) $\cong 7$ at $T - T_g = 3$ K and, hence, $\varepsilon''(\omega)$ will emphasize segmental motions of the PMPS block.

Results and discussion

The well-known Kohlrausch–Williams–Watt (KWW) function

$$C(t) = a \exp[-(t/\tau^*)^\beta] , \qquad (1)$$

was first employed to represent the experimental $C(t)$. Apart from the observed systematic deviations the shape parameter β was found to decrease from 0.29 at 348 K to 0.13 at 308 K thus reflecting

the broadening of $C(t)$. An alternative description, with no a priori assumption of the distribution but assuming a superposition of exponentials, is the inverse Laplace transformation of $C(t)$

$$C(t) = \int_{-\infty}^{\infty} L(\ln\tau) \exp(-t/\tau) \, d\ln\tau , \qquad (2)$$

$L(\ln\tau)$ denoting the distribution of retardation times is shown in Fig. 1 (lower part). The amplitude $a = \int_{-\infty}^{\infty} L(\log\tau) \, d\log\tau$ amounts to 0.7 at 308 K. Therefore, we conclude that a further fast fluctuating anisotropic intensity exists which becomes apparent in an asymmetric broadening of $L(\ln \tau)$ with decreasing temperature.

Figure 2 (upper part) depicts the frequency dependence of the dielectric loss ε'' for PS–PMPS at 305 K. A single Havriliak–Negami (HN) fit cannot adequately describe the observed $\varepsilon''(\omega)$ because of the asymmetric shape of $\varepsilon''(\omega)$ at low frequencies taken as a hint to an additional slow process. Strong evidence of a hidden slow process in Fig. 2 (upper part) is provided by plotting ε'' versus T in Fig. 2 (lower part). The value of ε''_{max} for the weak shoulder at high T is larger then expected for a fraction of PS in the diblock. This is further supported by a ratio of the relaxation strengths obtained from the fit of the experimental $\varepsilon''(\omega)$ to two HN functions which equals to 3. Therefore, it is unlikely that the slower relaxation is due to segmental orientation of the pure PS.

It is immediately apparent from Fig. 2 that the relaxation time $\tau_1 = 1/(2\pi\nu_{max})$ (ν_{max} being the frequency at the maximum dielectric loss for the fast process) is about seven decades faster than the correlation time τ_2 corresponding to the maximum of $L(\ln \tau)$ in Fig. 1 (lower part). Thus, the thermally and rheologically homogeneous PS–PMPS exhibits two separated primary relaxations which might be due to PS-rich and PMPS-rich local environments reflecting a bimodal composition fluctuation picture. The time scale $1/(2\pi\nu)$ for the slow process compares favorably with the τ_2 obtained from the PCS experiment. The disparity between τ_1 and τ_2 might arise from large $\langle |\delta f|^2 \rangle$. In this context, however, it is worth mentioning that an athermal ($\chi \cong 0$) diblock copolymer shows a bimodal distribution $L(\ln \tau)$ but with smaller separation [8].

In contrast, a microphase separated diblock copolymer exhibits two well-separated primary relaxations comparable in relaxation time and T dependence to the corresponding homopolymers [9]. Full description, analysis and discussion of all data on PS–PMPS block copolymer will be reported elsewhere.

Acknowledgement

The authors thank Mr. J. Gabinski for technical assistance.

References

1. Bates F, Fredrickson GH (1990) Annu Rev Phys Chem 41:525
2. Fredrickson GH, Helfand E (1987) J Chem Phys 87(1):697
3. Gerharz B (1991) Ph.D. Thesis, Mainz
4. Meier G, Fytas G (1989) In: Bässler H (ed.) Optical Techniques to Characterize Polymer Systems. Elsevier, Amsterdam
5. Saiz E, Floudas G, Fytas G, Macromolecules, in press
6. Fytas G, Patkowski A, Meier G, Fischer EW (1988) Macromolecules 21:3250
7. Kremer F, Boese D, Meier G, Fischer EW (1989) Progress in Coll Polym Sci 80:129
8. Kanetakis J, Fytas G, Kremer F, Pakula T, Macromolecules, submitted
9. Gerharz B, Fytas G, Fischer EW, Polym Commun, in press

Received January 21, 1992;
accepted May 29, 1992

Authors' address:

S. Vogt
Max-Planck-Institut für Polymerforschung
P.O. Box 3148
D-W-6500 Mainz, FRG

Progress in Colloid & Polymer Science   Progr Colloid Polym Sci 91:61–65 (1993)

Static and dynamic scattering at the microphase separation transition in block copolymers

A. Hoffmann, T. Koch, M. Schuler, F. Stickel, and B. Stühn

Fakultät für Physik, Universität Freiburg, FRG

Abstract: A series of polystyrene/(*cis* 1-4)polyisoprene diblock copolymers is studied using small-angle x-ray scattering (SAXS), real-time SAXS, photon correlation spectroscopy (PCS) and dielectric spectroscopy (DS). The static structure of concentration fluctuations is found to deviate from the Gaussian coil assumption in a temperature regime above the microphase separation transition (MST). The dynamic experiments in this temperature regime test concentration fluctuations (real-time SAXS) as well as single-block dynamics (PCS and DS) making use of the intrinsic labelling properties of this specific diblock copolymer. They provide evidence for the existence of partially ordered domains above the MST. The assumption of a weakly segregated state is, therefore, questionable.

Key words: Small-angle x-ray scattering – photon correlation spectroscopy – dielectric spectroscopy – block copolymers – microphase separation transition

Introduction

The microphase separation transition (MST) in block copolymers is considered as a disorder-to-order transition. It is, in this respect, similar to the crystallization of a liquid, the transition from paramagnetic into the antiferromagnetic state, and many other phenomena in physics. There are, however, certain peculiarities connected to this specific problem which arise from the polymeric nature of the system.

Polymer coils in the melt interpenetrate each other and, therefore, average intermolecular interactions over rather large scale. Their properties are, consequently, likely to be well described by mean field theories. Such a theory for the disordered state of diblock copolymers, called the random-phase approximation (RPA) [1–3] has provided a basis for the interpretation of many experiments on such findings [4]. There is, however, evidence from both experiments [5] and theory [6] that the assumption of the RPA concerning the Gaussian coil nature of the polymer conformation is not valid.

A further typically polymeric property that comes into play at the MST is the complex dynamics of long molecules. Whereas there are several studies on the static properties, the segmental dynamics of block copolymers was investigated in very few papers [7–9]. The comparatively slow dynamics of polymers makes it possible to use a variety of methods for the investigation of segmental dynamics. Furthermore, the dependence of local mobility on composition makes the segmental dynamics sensitive to concentration fluctuations. In some cases this coupling may lead to an interrelation between the MST and the glass transition [10].

In this contribution we will focus attention on the segmental dynamics of diblock copolymers at the MST. We use a series of diblocks of polystyrene/*b*-poly(*cis* 1-4)isoprene. Their molecular weights and compositions are given in Table 1. The interpretation of dynamic experiments is based on a characterization of the phase state of the samples with small-angle X-ray scattering (SAXS). The dynamic properties are then investigated using three different methods.

Table 1. Characterization of the diblock copolymers used

No.	Polystyrene volume fraction	M_W	M_W/M_N	T_s (K)	T_t (K)	T_{MST} (K)
1	0.50	10 500	1.04	336	356	—[a]
2	0.44	15 700	1.04	391	403	362
3	0.77	20 800	1.06	343	355	—[a]

[a]) Not observed

The dynamics of concentration fluctuations are observed in temperature quench experiments using real-time SAXS. The single block dynamics may be studied by making use of the intrinsic labelling properties of the polystyrene/polyisoprene diblocks. In depolarized light scattering (photon correlation spectroscopy: PCS) the polystyrene block is essentially labelled because of its strong optical anisotropy. On the other hand, polyisoprene provides an in-chain component of the segmental dipole moment which gives rise to a normal mode in the dielectric spectrum.

We will show that the combination of these methods results in a revised picture of the disordered state in diblock copolymer melts.

Results

The purpose of the present contribution is a survey of results from several experimental techniques. For a detailed discussion of the experiments we will, therefore, refer to more specialized treatments.

SAXS has been applied for the investigation of the disordered state in polystyrene/polyisoprene diblock copolymers [7, 5, 10–12]. The profiles exhibit a broad peak at a characteristic scattering vector q^*. Data analysis is normally based on RPA theory and results in the temperature dependence of the interaction parameter χ and a spinodal temperature T_s. Both quantities govern the size of concentration fluctuations in the disordered state. With lowering temperature, clear deviations from RPA theory are visible in the SAXS profiles, in particular, the linear dependence of the reciprocal peak intensity $I(q^*)^{-1}$ on reciprocal temperature does not hold for temperatures $T \leq T_t$. This temperature marks the begin of a transition regime which is no longer described by simple RPA theory. It is characterized by the presence of strong concen-

tration fluctuations [2] which also lead to a modification of the disorder-to-order transition as expected by RPA. T_s and T_t, therefore, must be considered phenomenological parameters. Sample 2 is unique in clearly displaying the MST as the formation of long-range order. In the SAXS experiments one finds a discontinuity in the peak intensity and the appearance of a second-order reflection [13]. In both other examples the beginning of structure formation is stopped by the glass transition of the polystyrene domain [10]. All characteristic temperatures are listed in Table 1.

The stong change in peak intensity between the disordered state and temperatures in the transition regime allows the measurement of the time evolution of the SAXS intensity in a temperature jump experiment. Sample 3 is annealed at $T_{anneal} = 413$ K and then quenched below the glass transition of the disordered phase. The success of the quench is checked by a comparison of the SAXS profiles at 413 K and in the quenched state. The sample is then pushed into the oven of the x-ray camera. Within less than 7 s it has reached the final temperature T. The effective temperature program, therefore, is a jump $T_{anneal} \rightarrow T$.

The time dependence of the x-ray intensity at scattering vector q is directly related to the correlation function of concentration fluctuations with wave vector q [14]:

$$I(q, t) \propto \langle \Phi_q(0)\,\Phi_q(t) \rangle \,. \tag{1}$$

The experiment averages over a q interval $\Delta q \approx 0.2$ nm^{-1} around q^*. We, therefore, restrict our attention to the time dependence of $I_{q*}(t)$. Figure 1 shows the result of $I_{q*}(t)$ for a temperature jump 413 K $\rightarrow$ 343 K. It is clearly seen that the observed relaxation cannot be described by a single exponential. It is composed of two distinct modes:

$$I_{q*}(t) = I(0) + \Delta I_1\{1 - \exp(-t/\tau_1)\}$$
$$+ \Delta I_2\{1 - \exp[-(t/\tau_2)^\beta]\} \,. \tag{2}$$

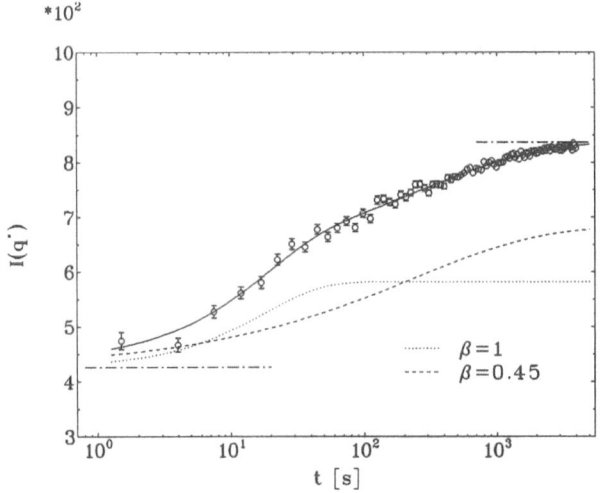

Fig. 1. $I_{q^*}(t)$ for sample 3 at a T jump 413 K → 343 K. The dotted lines show the decomposition into two modes

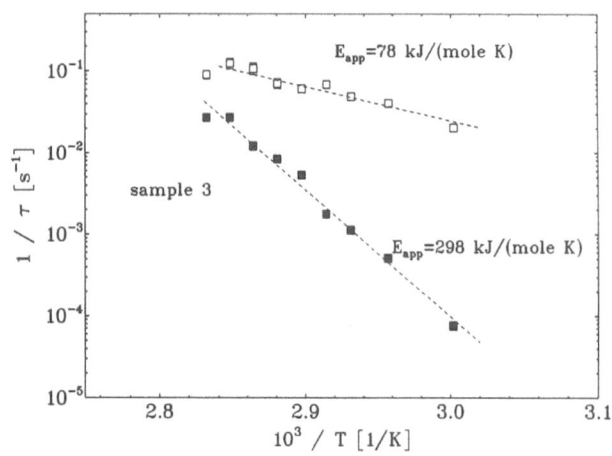

Fig. 2. Temperature dependence of the relaxation times τ observed in the real-time SAXS experiment

At short times (τ_1) one finds a single exponential whereas the relaxation at long times is of the Kohlrausch–Williams–Watts (KWW) type with an exponent $\beta = 0.45$. Figure 2 displays the temperature dependence of the relaxation times τ_1 and τ_2 in an Arrhenius representation. Clearly, different apparent activation energies suggest τ_1 to correspond to the formation of local order or concentration fluctuations within the mobile, mixed phase. The second increase in intensity, τ_2, is caused by the building of long-range order. This involves the motion of polystyrene-rich domains and is, therefore, coupled to their glass transition.

The second dynamic experiment aims at the mobility of the polystyrene block [15]. It is known that polystyrene gives rise to strong depolarized light scattering which may be used to study the relaxation of orientational correlations in a PCS experiment [16]. Above the glass transition (374 K to 402 K) one obtains relaxation curves $g_1(t)$ of the KWW type with β varying from 0.4 to 0.3. The temperature dependence of the relaxation curves is well described within the WLF theory [17].

The correlation functions obtained on the symmetric diblock (sample 1) are shown in Fig. 3. The temperature range corresponds to the transition regime. The stretching of the curves is even more pronounced than for bulk polystyrene. Figure 4 shows in more detail our model for a fit of the relaxation curves. The KWW exponents β, how-

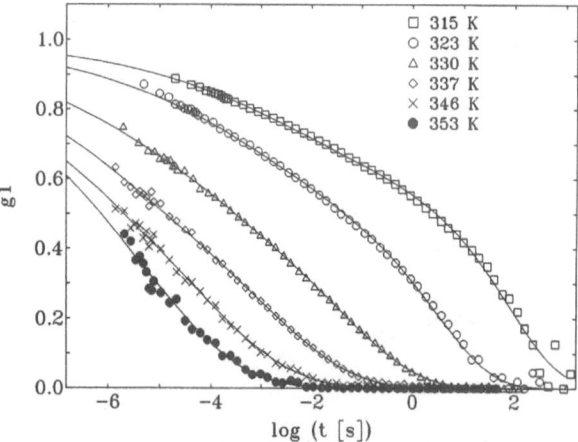

Fig. 3. PCS results of sample 1. The drawn lines are fits using two KWW functions

ever, are assigned fixed values $\beta_{slow} = 0.4$ and $\beta_{fast} = 0.2$ for the fast and the slow relaxation, respectively. They correspond to bulk polystyrene and polystyrene mixed with isoprene as was shown in separate measurements [15]. Figure 5 now displays the temperature dependence of both processes together with the results of bulk polystyrene. The slow relaxation seems to emerge from the polystyrene data essentially through a shift of the glass transition temperature. The full curve in Fig. 4 is calculated using free-volume theory [18] and assuming the polystyrene to be mixed with 18% polyisoprene.

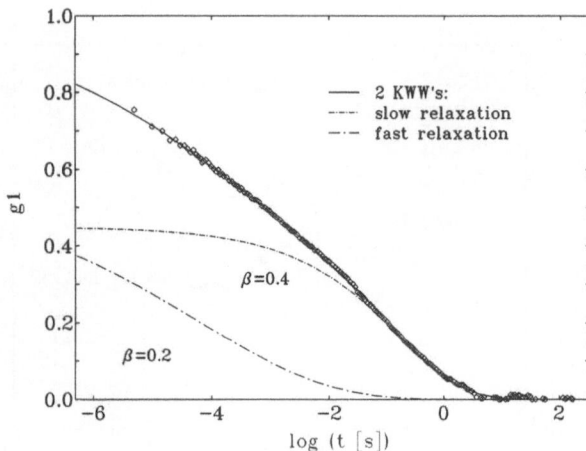

Fig. 4. Decomposition of the correlation function into two KWW's

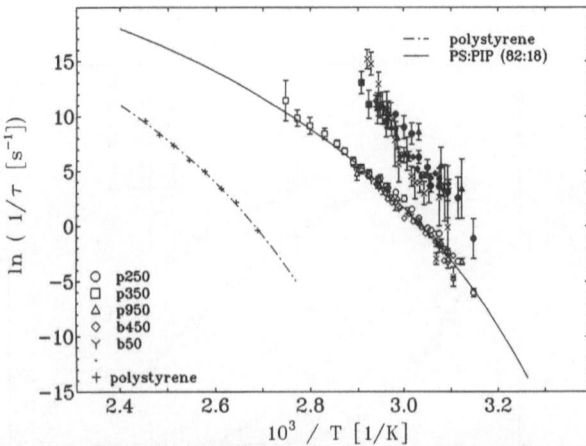

Fig. 5. Temperature dependence of the average relaxation times observed in PCS on polystyrene and the symmetric diblock copolymer (sample 1). The different symbols for the block copolymer results refer to repeated measurements under identical conditions

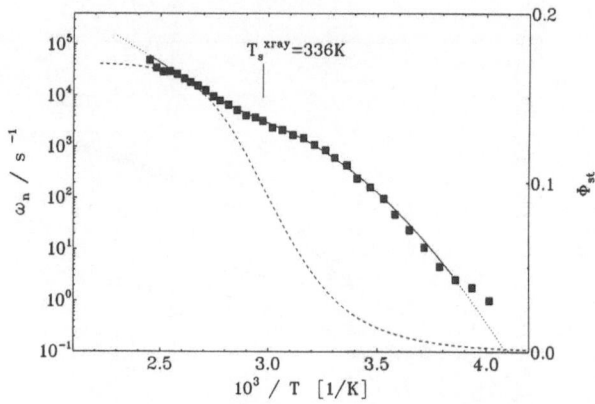

Fig. 6. Temperature dependence of the normal-mode frequency for sample 1 (left scale). The broken line shows the variation of the polystyrene volume fraction Φ_{St} (see text) in the polyisoprene domains (right scale). The resulting $\omega_n(T)$ is included as the full curve

This description is not applicable to the fast mode. Its temperature dependence suggests a variation of composition with T. With lowering temperature there is a reorganization in the highly mobile, polyisoprene-rich domains by further expulsion of polystyrene. The polystyrene-rich phase grows at constant composition.

Complementary to the PCS experiment, one obtains information on the dynamics of the polyisoprene block from measurements of the dielectric relaxation [19]. The intrinsic labelling is in this case

accomplished by the existence of a normal mode which is unique for the polyisoprene block. It is caused by an in-chain dipole moment per unit length [19] $\tilde{\mu}_{||} = 0.45$ D/nm of isoprene. The relaxation of the end-to-end vector is observed as a broad mode in the dielectric permittivity $\varepsilon''(\omega)$ (see Fig. 6). The characteristics of this mode are well studied in bulk polyisoprene [20, 21]. Here we only discuss the temperature dependence of the relaxation frequency $\omega_n = 1/\tau_n$. In the Arrhenius plot (Fig. 6) one observes a peculiar dip around T_s. At low temperatures, in the ordered state, ω_n follows the WLF curve known for polyisoprene [20]. Around T_s the environment of the polyisoprene block becomes enriched with polystyrene and the effective friction coefficient increases. The full curve in Fig. 7 is calculated for polyisoprene with a volume fraction $\Phi_{St}(T)$ polystyrene (see above). The temperature variation of Φ_{St} is assumed to be of tanh type and also given in the figure. In a temperature interval of ≈ 30 K around T_s the polyisoprene domains become mixed with polystyrene. The mixing does not lead into the homogeneous, disordered state as is seen in Fig. 6: $\Phi_{St}(T)$ approaches a value much smaller than the global composition f.

Conclusions

Using a combination of static and dynamic experiments we have studied the properties of diblock copolymers in the transition regime close to the MST. The structure of the polymer coils is seen to deviate significantly from the Gaussian conformation. Time-resolved SAXS experiments during a quench below T_s show the formation of local and long-range order on separate time scales. The local ordering is present in the transition regime. It results in concentration fluctuations which have been characterized using PCS and dielectric spectroscopy on the segmental scale as well as on the scale of the polyisoprene block. The MST leads to changes of the local concentration only on the scale of a block and not on the segmental scale. The MST can, therefore, not be described by theories based on the assumption of a weakly segregated state.

Acknowledgements

We are grateful to Prof. Strobl for helpful discussions.

References

1. Leibler L (1980) Macromolecules 13:1602
2. Fredrickson GH, Helfand E (1987) J Chem Phys 87:697
3. Binder K, Fredrickson GH (1990) J Chem Phys 10:6195
4. Bates FS, Fredrickson GH (1990) Ann Rev Phys Chem 41:525
5. Holzer B, Lehmann A, Stühn B, Kowalski M (1991) Polymer 32:1935
6. Fried H, Binder K (1991) Europhys Lett (1991) 16:237
7. Stühn B, Rennie AR (1989) Macromolecules 22:2460
8. Kanetakis J, Fytas G, Hadjichristidis N (1991) Macromolecules 24:1806
9. Hashimoto T (1987) Macromolecules 20:465
10. Stühn B (1992) J Polym Sci Polym Phys Ed 30:1013
11. Owens JN, Gancarz IS, Koberstein JT, Russel TP (1989) Macromolecules 22:3380
12. Mori K, Tanaka H, Hasegawa H, Hashimoto T (1989) Polymer 30:1389
13. Stühn B, Mutter R, Albrecht T (1992) Europhys Lett 18:427
14. Binder K (1983) J Chem Phys 79:6387
15. Hoffmann A, Koch T, Stühn B, to be published
16. Patterson CP, Stevens JR (1979) J Polym Sci 17:1547
17. Koch T (1991) PhD Thesis, Universität Freiburg
18. Fujita H (1961) Fortschr Hochpolym Forsch 3:1
19. Stühn B, Stickel F (1992) Macromolecules, in press
20. Adachi K, Kotaka T (1985) Macromolecules 18:466
21. Boese D, Kremer F (1990) Macromolecules 23:829

Received December 17, 1991;
accepted May 24, 1992

Authors' address:

Dr. B. Stühn
Universität Freiburg
Fakultät für Physik
Hermann-Herder-Str. 3
D-W-7800 Freiburg, FRG

Progress in Colloid & Polymer Science   Progr Colloid Polym Sci 91:66–68 (1993)

Mode coupling corrections to the Onsager coefficient as determined by light scattering of critical concentration fluctuations from polymer mixtures

G. Meier, B. Momper, and E. W. Fischer

Max-Planck-Institut für Polymerforschung, Mainz, FRG

Abstract: We report static and dynamic light-scattering experiments of an almost symmetric polymer mixture made up from polydimethylsiloxane (PDMS), $N = 260$ and polyethylmethylsiloxane (PEMS), $N = 340$ with N being the degree of polymerization. The mixture exhibits an upper critical solution temperature $T_c \simeq 57\,°C$ with a critical composition $\phi_{C,\,PEMS} = 0.465$. The measurements were performed in a broad temperature range in the one-phase region between $-0.7 < \varepsilon < -2.5$ with $\varepsilon = T - T_c/T$. From measurements of the mutual diffusion coefficient $\tilde{D}$ and the static structure factor $S(q = 0)$, the Onsager coefficient $L(q) = \tilde{D}S(q = 0)$ was calculated. It is related to W^0, a microscopic frequency, given by the Rouse diffusion coefficient $D_R^0 = W^0/N$ via $L = \phi(1 - \phi)W^0$. We have basically found that for values of the correlation length ξ of concentration fluctuations smaller than the coil size, W^0 behaves according to Rouse theory, but in the hydrodynamic regime, where one probes the slow diffusive dynamics of concentration fluctuations with wavelengths larger than the size of the polymer coil and the correlation length, W^0 tends to accelerate as $T \to T_c$. Using the mode–mode coupling prediction that in the cited regime W^0 is proportional to ξ we have removed the apparent discrepancy.

Key words: Critical phenomena – light scattering – polymer mixtures – Onsager coefficient – mode–mode coupling

As a system approaches the critical point, the fluctuations of the order parameter become very large as a consequence of the divergence of the generalized susceptibility of the system. We consider the case of a binary polymer mixture at the critical composition (PEMS, $N = 340$, $\phi_c = 0.465$; PDMS, $N = 260$). Here the order parameter is given by $\phi - \phi_c$, its susceptibility is given by $(\partial\phi/\partial\Delta)_{T,P}$, where $\Delta = \mu_1 - \mu_2$ is the difference between the chemical potentials of the two components. We have measured the central, quasielastic component in the spectrum of scattered light which is caused by the diffusive decay of the concentration fluctuations. Its decay rate is given by [1, 2]

$$\Gamma(q) = \frac{L(q)}{S(q)}q^2 \,, \qquad (1)$$

where $L(q)$ is the Onsager coefficient and $S(q)$ is the generalized susceptibility. As T approaches T_c, S diverges and, hence, Γ has to go to zero, commonly known as critical slowing down [1]. At the same time the Onsager coefficient is assumed to vary slowly in the critical region. Some critical coefficient, however, may change rapidly near T_c due to nonlinear couplings between the hydrodynamic modes [3–6].

We were interested in the evaluation of L over a wide temperature range which at a given value of q, represents a change of the correlation length ξ for values of ξ smaller than the coil size (called region I) up to a regime where q^{-1} is much bigger than the size of the polymer coil and ξ. This regime is also a hydrodynamic regime (region II) [7]. In principle, there is a further thrid regime where ξ is much larger than q^{-1} but this is difficult to reach.

Let us rewrite Eq. (1), thereby assuming an Ornstein–Zernicke scattering law for $S(q)$:

$$\Gamma(q) = L(q)q^2\left[\frac{1 + q^2\xi^2}{S(q = 0)}\right].$$ (2)

For $q\xi \ll 1$, this reduces to $\Gamma(q) \equiv Dq^2 = L(q)S(q=0)^{-1}$, a result which is valid for region I [8]. This was proven in this region by plotting $\log\Gamma$ vs. $\log q$ yielding a slope of 2 within the experimental accuracy. From the dynamic random-phase approximation [7, 9], we know that for $q\xi \ll 1$, $L = \phi(1 - \phi)W^0$; hence, the molecular weight and wave vector independent local (microscopic) frequency $W^0 = (N/a^2)D_R^0$ with D_R^0 being the Rouse diffusion coefficient, can be calculated. The characteristic length a in the mixture is defined by $a^2 = (1 - \phi)\sigma_A^2 + \phi\sigma_B^2$ with $\sigma_{A,B}$ being related to the end to end distance R of the polymer coils by $R^2 = \sigma^2 N$. For Gauß coils $6R_g^2 = R^2$ with R_g being the radius of gyration. The values for W^0 as obtained by measuring $\Gamma(q)$ and $S(q = 0)$ are plotted as a function of the inverse temperature in Fig. 1. The region I is valid for temperatures roughly above $T = 80\,°C$. However, if we lower T, mode–mode coupling corrections become obviously necessary (region II). In critical dynamics the measured linewidth Γ is usually given by $\Gamma = \Gamma^C + \Gamma^B$, where Γ^C is the critical part and Γ^B the background part of the linewidth [2]. It has been shown that for polymer mixtures $\Gamma^B \ll \Gamma^C$ [10, 12], so we assume that the measured linewidth for $T < 80\,°C$ is mainly due to the critical part Γ^C

of Γ ($\Gamma^C \approx \Gamma$) [10]; then

$$\Gamma^C \simeq \phi(1 - \phi)\frac{W^0q^2}{S(q = 0)}F(q\xi),$$ (3)

where $F(q\xi)$ is a dimensionless function which results from expanding the Kawasaki function in region II. Here, $\Gamma^C/q^3 = kT/6\pi\eta(q\xi)^{-1}F(q\xi)$ [4, 7]; hence, $D^C \equiv \Gamma^C q^{-2} \propto \xi^{-1}$. The square of the static correlation length ξ is given by the slope of $S(q = 0)/S(q)$ versus q^2 (Zimm plot), which implies that further $S(q = 0) \propto \xi^2$ [11]. We conclude from the former proportionalities that $W^0 \propto \xi$ (valid in region II). This is the essential prediction which we have tested in Fig. 2. Indeed, $W^0\xi^{-1}$ vs. the reduced temperature $\varepsilon = T - T_c/T$ remains constant from a value of $\xi(T)$, which corresponds to lengths of the order of R_g and larger. The temperature where the value of ξ reaches the coil size occurs at a temperature corresponding to $\log\varepsilon = -1.1$. This corresponding temperature is indicated by an arrow in Fig. 1. From these findings we conclude that the deviations from a straight line in Fig. 1 (the straight line can be calculated from mechanical measurements assuming the Rouse model [10]) for $T < 80\,°C$ can be explained by using mode–mode coupled expressions which, of course, are not in the scope of the random-phase approximation [7]. Finally, it is

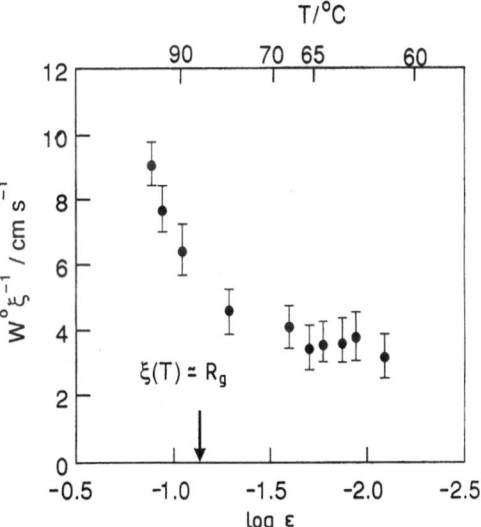

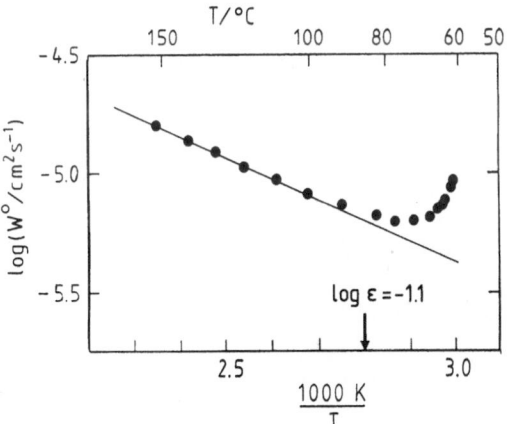

Fig. 1. Arrhenius plot of the microscopic frequency W^0. At $T \approx 80\,°C$ (arrow) the value of the correlation length $\xi(T)$ corresponds to the coil size

Fig. 2. The ratio of microscopic frequency W^0 to the correlation length ξ versus the reduced temperature ε. For values of ξ larger than R_g, the ratio remains constant which means that mode–mode coupled dynamics come into play

important to state that Stepanek et al. [12] have also drawn a similar conclusion from their experiments; namely that the coil size is the relevant length to distinguish between regions I and II and hence, between nonmode–mode coupled and mode-mode coupled dynamics.

References

1. van Hove L (1954) Phys Rev 95:1374
2. Swinney H, Henry D (1973) Phys Rev A 8:2586
3. Fixman M (1962) J Chem Phys 36:310
4. Kadanoff LP, Swift J (1968) Phys Rev 166:89
5. Oxtoby D, Gelbart W (1973) J Chem Phys 61:2957
6. Kawasaki K (1970) Ann Phys (NJ) 61:1
7. Fredrickson GH (1986) J Chem Phys 85:3556
8. Brereton MG, Fischer EW, Fytas G, Murschall U (1987) J Chem Phys 86:5174
9. Brochard F, de Gennes PG (1983) Physica A 118:289
10. For a detailed analysis see the extended paper: Meier G, Momper D, Fischer EW (1992) J Chem Phys 97:0000
11. Binder K (1983) J Chem Phys 79:6387
12. Stepanek P, Lodge TP, Kedrowski C, Bates FS (1991) J Chem Phys 94:8289

Received January 24, 1992;
accepted June 5, 1992

Authors' address:

G. Meier
KFA Jülich
IFF Postfach 1913
D-W-5170 Jülich, FRG

PEO–PPO–PEO block polymer in aqueous solution: Micelle formation and crystallization

K. Mortensen

Physics Department, Risø National Laboratory, Roskilde, Denmark

Abstract: The structural properties of PEO–PPO–PEO block copolymers dissolved in water have been studied by small-angle neutron scattering. Three different phases appear. At low temperature ($T \leq 15\,°C$) the block copolymers are dissolved as individual Gaussian chains (unimers). At intermediate temperatures, the hydrophobic nature of PPO causes aggregation into a liquid of micelles. At even higher temperature ($T \geq 36.5\,°C$ for the 28% polymer solution) the micellar liquid "freezes" into a cubic crystalline powder, which can be aligned by shear into a single crystal.

Key words: Block copolymers – micelle – colloidal crystal – cubatic

Introduction

Aqueous solutions of poly(ethylene oxide)–poly(propylene oxide) triblock copolymers, PEO–PPO–PEO, exhibit dramatic structural and dynamical dependence on temperature, mainly governed by the hydrophilic-to-hydrophobic transition in PPO. Based on dynamic light scattering, Zhou and Chu reported in 1987 that micelles are formed above a critical concentration and temperature [1]. PEO–PPO–PEO is moreover known to undergo a reverse thermal gelation, as seen by rheological methods [2–5]. A variety of suggestions have been proposed for this sol–gel transition, including entropic changes due to locally ordered water molecules, percolation threshold of micelle rods, or perhaps some local three-dimensional ordering.

In a recent publication, we have shown that at least three distinct phases of the poloxamer solution appear, as the temperature is varied [6]: a low-temperature dispersion of independent triblock units, an intermediate temperature micellar liquid, and a high-temperature micellar crystal.

Results and discussion

In the present letter, we report structural studies on a 28% solution of the 10 800 molecular weight

F88 poloxamer ($EO_{97}PO_{39}EO_{97}$) dissolved in deuterated water. Small-angle neutron scattering was done using the Risø-SANS facility. The sample was mounted in various holders, including a Couette type of shear cell [7]. Deuterated water, D_2O, was used in order to get good contrast between the polymer and the solvent, and low background in the scattering experiments. The contrast between PEO and PPO is negligible.

At low temperature, i.e., below approximately $10\,°C$, the scattering function is that of a Gaussian chain, with radius of gyration of the order of 20 Å; thus, indicating independent PEO–PPO–PEO linear polymer coils.

At temperatures above $15\,°C$ the intensity of the scattered beam increases markedly, revealing that PEO–PPO–PEO aggregates because water in this temperature regime acts thermodynamically as a precipitant for the PPO-block. Above $20\,°C$, the scattering function is dominated by a pronounced correlation peak. Figure 1 shows a typical example of $I(q)$ data as obtained in this regime. The scattering function,

$$I(q) = NKS(q)P(q) , \qquad (1)$$

has been analyzed in terms of the Percus–Yevick approximation [8] using a hard-sphere interaction potential. The structure factor, $S(q)$, is then given by the analytical form

$$S(q) = \frac{1}{1 + 24\phi G(2qR_{hs}, \phi)/(2qR_{hs})} , \qquad (2)$$

where ϕ is the hard-sphere volume fraction, and G is a function given by ϕ and trigonometric functions of $2qR_{hs}$, R_{hs} being the hard-sphere interac-

tion radius. The formfactor, $P(q)$, is assumed to be dominated by the dense spherical core:

$$P(q) = \frac{3}{(qR_c)^3} [\sin(qR_c) - qR_c \cos(qR_c)] , \quad (3)$$

R_c being the core radius. N is the number of scatters and K is the contrast factor. In the least-square fitting routine, we include instrumental smearing effects. Figure 1 includes the fit as well as the scattering function with no smearing.

In the temperature regime from 15 °C to 36.5 °C, the Percus–Yevick approximation gives excellent fits, while the correlation peak above $T_m = 36.5$ °C becomes narrower than the PY model allows. Figure 2 shows the volume fraction, ϕ, resulting from the least-squares fit. It is seen that the volume fraction of micelles increases linearly, until ϕ reaches a critical, limiting value where the system locks-in to a crystalline state.

If a small shear field is applied to the 28% aqueous F88 solution at temperatures above $T_m = 36.5$ °C, the azimuthally isotropic scattering pattern abruptly transforms into a hexagonal pattern of Bragg reflections; thus, revealing the formation of crystalline phase. In Fig. 3 is shown the two-dimensional scattering pattern of 28% F88 as obtained just above the "inverse melting transition" at $T_m = 36.5$ °C.

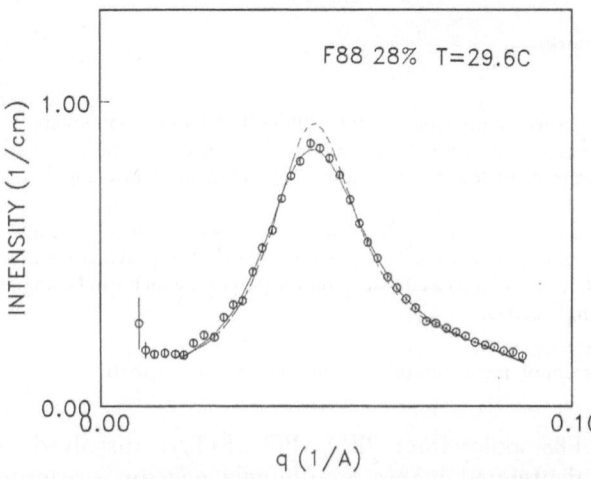

Fig. 1. Scattering function of 28% aqueous solution of F88, as obtained at $T = 29.6$ °C. The solid line represents fit using the hard-sphere approximation, Eqs. (1)–(3). The dashed curve represents the same function with no instrumental smearing

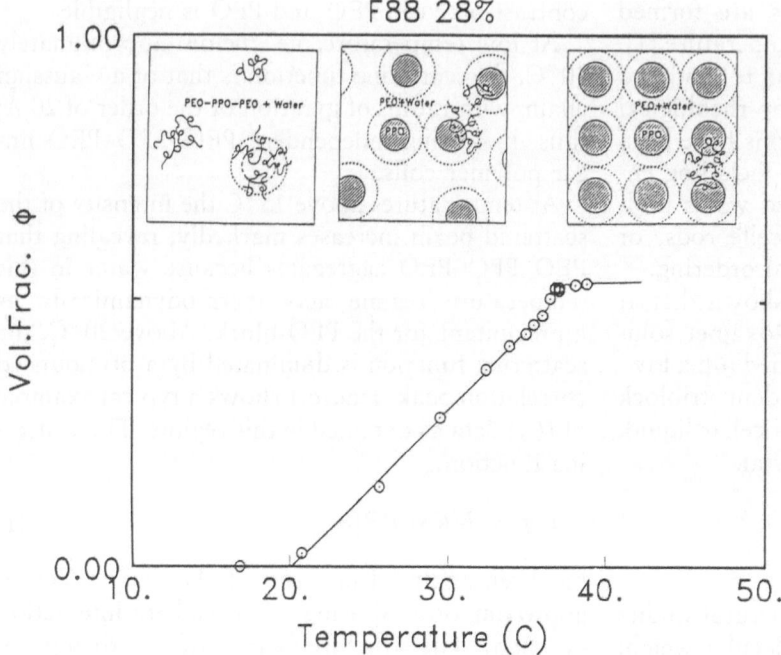

Fig. 2. Micellar volume fraction, as obtained by Percus–Yevick fits to the scattering data of 28% F88. The insert shows the structure of F88 in the three temperature ranges: low-temperature unimers in solution, intermediate temperature micellar liquid, and high-temperature cubic colloidal crystal

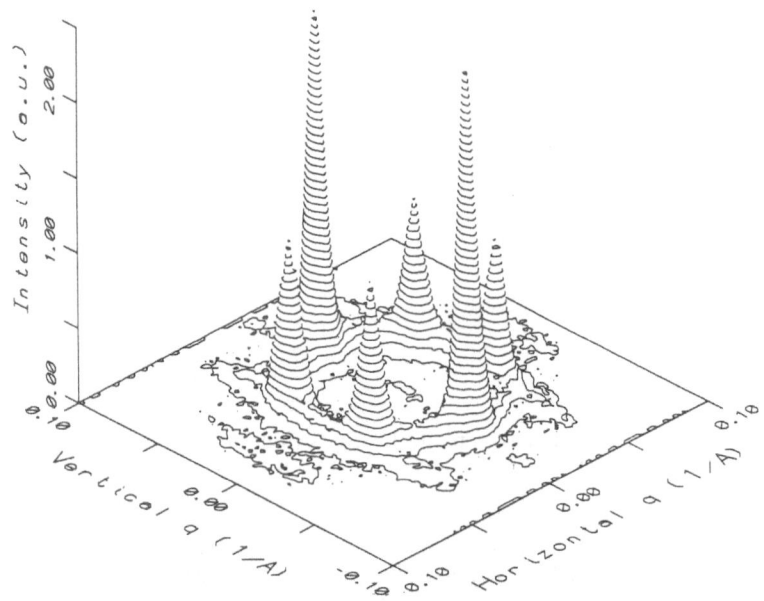

Fig. 3. Two-dimensional scattering function of shear oriented 28% F88, as obtained at 38 °C, showing the [111]-plane of the BCC-structure [6]

The (110)-Bragg reflections seen in Fig. 3 are not resolution limited. Fitting to the Lorentz function: $I(q) = I(0)/[1 + (q\xi)^2]$, and including instrumental resolution, results in a bond correlation length, ξ, of only 600 Å. Still the angular correlation length, as seen by the well-resolved Bragg reflections in Fig. 3, covers the whole sample, i.e., is of the order of centimetres. These findings lead to the conclusion that the colloidal crystal is a "cubatic" liquid crystal in which the bond-angle correlation is the relevant order parameter [6].

In summary we have shown that the 28% aqueous solutions of F88 ($EO_{97}PO_{39}EO_{39}$) block copolymers have three characteristic temperature regimes, as displayed in the insert of Fig. 2. At low temperature, the polymers are dissolved as independent coils and only minor aggregation appears. At intermediate temperatures, the hydrophobic interactions cause micelle formation

$$unimers \rightleftharpoons micelles$$

with T-dependent equilibrium constant. As the temperature is raised, both the number of micelles and the micelle sizes increase, resulting in an increasing volume fraction. The sol–gel transition, as observed in rheological studies, appears as a result of reaching the critical volume fraction, causing a lock-in into a hard-sphere colloidal crystal.

Acknowledgements

The present study is a part of an extended study on the structure and dynamics of poloxamer in aqueous solutions. I greatly acknowledge illuminating discussions with my coworkers in this field: Wyn Brown (Uppsala University), J. Skov Pedersen (Risø National Laboratory), and S. Hvidt (Roskilde University). I thank Bengt Nordén (Chalmers Institute of Technology) for using his Couette shear cell. Financial support from the Danish Natural Science Research Council is gratefully acknowledged.

References

1. Zhou Z, Chu B (1987) Macromolecules 20:3089; (1988) J Colloid Interface Sci 126:171
2. Vadnere M, Amidon GL, Lindenbaum S, Haslam JL (1984) Int J Pharm 22:207
3. Wanka G, Hoffmann H, Ulbricht W (1990) Colloid Polym Sci 268:101
4. Brown B, Schillen K, Almgren M, Hvidt S, Bahadur P (1991) J Phys Chem 95:1850
5. Wang P, Johnston TP (1991) J Appl Polymer Sci 43:283
6. Mortensen K, Brown W, Nordén B (1992) Phys Rev Letters 13:2340
7. Nordén B, Elvingson C, Eriksson T, Kubista M, Sjöberg B, Takahashi M, Mortensen K (1990) J Mol Biol 216:223
8. Kinning DJ, Thomas EL (1984) Macromolecules 17:1712

Received November 7, 1991;
accepted June 17, 1992

Author's address:

K. Mortensen
Physics Department
Risø National Laboratory
DK-4000 Roskilde, Denmark

Progress in Colloid & Polymer Science   Progr Colloid Polym Sci 91:72–74 (1993)

Dynamic light scattering study of a 1,4-isoprene-b-styrene copolymer

A. K. Rizos[1]), G. Fytas[1]), J. E. L. Roovers[2]), and K. L. Ngai[3])

[1]) Foundation for Research and Technology Hellas, Heraklion, Crete, Greece
[2]) National Research Council of Canada, Ottawa, Ontario
[3]) Naval Research Laboratory, Washington, DC, USA

Abstract: Local optical anisotropy fluctuations within polystyrene (PS)-rich region in a disordered 1,4-isoprene-b-styrene (PI-b-PS) copolymer with $\chi N \approx 5.3$ ($N = 44$) and glass transition temperature $T_g = 13\,°C$ were analyzed by photon correlation spectroscopy (PCS) over a broad time range (10^{-7}–10^3 s). There is strong experimental evidence that composition fluctuations in the present system can affect the segmental orientational fluctuations of the two foreign blocks. The significant broadening of the distribution of orientation times with decreasing temperature toward T_g can be rationalized in terms of the coupling scheme of relaxation.

Key words: Diblock copolymers – dynamic light scattering – concentration fluctuations – coupling model

Introduction

Block copolymers A-B have attracted considerable attention during the last few years both from the theoretical and experimental point of view [1–3]. The vast majority of the previous investigations focused on the static properties mainly in the ordered phase by means of X-ray or small-angle neutron scattering (SANS) and dynamic mechanical measurements. In contrast to that, only few dynamic investigations [4–7] have appeared in the literature. A central role in the theory of diblock copolymers play the local composition fluctuations $\langle \Psi_q^2 \rangle$, where Ψ_q is the Fourier transform of $\Psi(r) = \rho_A(r)/\rho - f$ with ρ_A, ρ and f being the number density of A, total of (A + B) density and composition of A in the copolymer. The amplitude $\langle \Psi_q^2 \rangle$ is predicted to increase by approaching the microphase separation transition (MST) and expected to affect the local segmental dynamics of the blocks.

The aim of this work is to use techniques that are sensitive to local motions like photon correlation spectroscopy (PCS) which covers a very broad dynamic range. Also by taking advantage of the selectivity of the technique by employing the so-called VH scattering geometry, we can probe selectively the dynamics of the segments with the highest anisotropy. We have chosen for this investigation the diblock polystyrene-1,4-polyisoprene (PI-b-PS) with low molecular weight $M_n(PS) = 2830$, $M_n(\text{total}) = 3930$, whose segment–segment interaction parameter is $\chi = -0.0937 + 66/T$ [8] (T is the temperature) and, hence, this sample with $\chi N \approx 5$ is in the homogeneous state above its glass transition temperature [9] $T_g(= 13\,°C)$.

Experimental results

The correlation functions $G(t)$ of the depolarized VH light scattering intensity were measured at different T near T_g at a scattering angle of $90°$ with an ALV-5000 multiple sampling time digital correlator over a wide time range (10^{-7}–10^3 s). The correlation function $C_{VH}(t)$ for the fluctuations in the optical anisotropy is then computed from $C_{VH}(t) = [G(t) - 1]^{1/2}$ since the sample displays only inherent light scattering. Figure 1 shows two representative $|C_{VH}(t)|^2$ for the

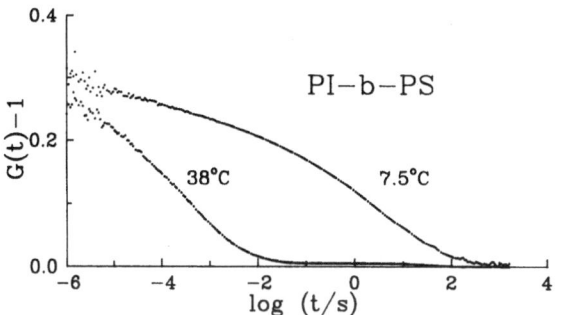

Fig. 1. Depolarized intensity correlation functions for the PI-b-PS diblock copolymer at 38 °C and 7.5 °C

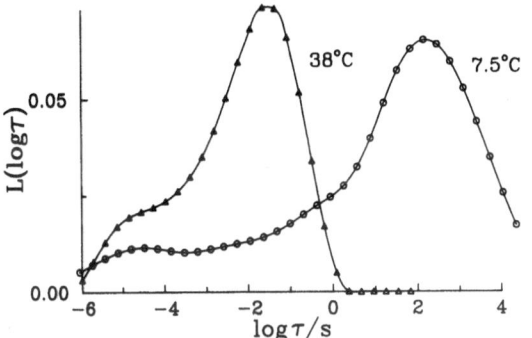

Fig 2. Distribution of retardation times obtained from the inverse Laplace transform analysis (ILT) of the experimental time correlation functions at two temperatures

PI-b-PS diblock at two temperatures above T_g. The functions $C_{VH}(t)$ have been fitted first to the Kohlrausch–Williams–Watts (KWW) expression $\exp(-t/\tau)^\beta$ yielding the values of the relaxation time τ and the distribution parameter β. Due to observed deviations to the KWW fit at temperatures close to the T_g, we have then carried out the inverse Laplace transformation (ILT) of $C_{VH}(t)$. Assuming a superposition of exponentials, the ILT of $C_{VH}(t)$ gives the continuous spectrum of orientation times $L(\log\tau)$ shown in Fig. 2.

Discussion

The amplitude of $C_{VH}(t)$ at short times is a measure of the fraction of the depolarized Rayleigh intensity arising from fluctuations with correlation times longer than 10^{-7} s. This value can be estimated, if the values of the monomeric optical anisotropies $\langle\langle\gamma^2\rangle\rangle$ of the copolymer

components are known. For the present PI-b-PS with $f_{PS} = 0.61$, $\langle\gamma^2\rangle_{PS} = 38\ \text{Å}^6$ and $\langle\gamma^2\rangle_{PI} = 11\ \text{Å}^6$, the function $C_{VH}(t)$ reflects mainly the contribution of segmental orientation of the PS block. The segmental dynamics of the PI block are faster than 10^{-7} s and have recently been studied by dielectric spectroscopy [12].

The experimental finding that the present disordered PI-b-PS sample displays two distinct, well-separated, segmental primary relaxations constitutes strong support of the presence of significant $\langle\Psi_q^2\rangle$ suggested by the fluctuation picture of diblock copolymers [2, 3]. Here, we consider the segmental dynamics within PS-rich region showing a strong T-dependent distribution of orientation times (Figs. 1 and 2) over the T-range 55–6 °C. Over this range, χN varies from 4.8 to 6.2.

The distribution of segmental relaxation times in the disordered state can be rationalized by the coupling model as generalized recently for miscible blends [10]. Due to concentration fluctuations $\langle\Psi_q^2\rangle$, the basic PS units in segmental orientational relaxation will have a distribution of local environments which in turn gives rise to a distribution of coupling parameters of segmental relaxation. Environments richer in neighboring PS chain units cause a larger coupling parameter n, because coupling between PS and PS is stronger than between PS and PI. In other words, local environments richer in PS confer to locally rearranging PS segments, a larger degree of intermolecular coupling and, hence, larger n. The effective relaxation times τ_i^*,

$$\tau_i^* = [(1 - n_i)\omega_c^{n_i}\tau_0(T)]^{1/(1-n_i)}, \tag{1}$$

where $\tau_0(T)$ the PS segmental relaxation times without considering the coupling to other chains, shift differently with temperature according to the size of n_i and broaden the relaxation spectrum (see Fig. 2). The τ_i^*'s corresponding to the largest n_i's shift more to longer times as T decreases, because of the nonlinear nature of the dependence of τ_i^* on n_i. Hence, the broadening is more prominent on the long-time side of the spectrum (see Fig. 2). The coupling model can bring out additional physics from the dynamic light-scattering data. The details will be reported elsewhere [11].

References

1. Leibler L (1980) Macromolecules 13:1602
2. Fredrickson GH, Helfand E (1987) J Chem Phys 87:697

3. Bates FS, Fredrickson GH (1990) Ann Rev Phys Chem 41:525
4. Stuhn B, Rennie AR (1989) Macromolecules 22:2460
5. Kanetakis J, Fytas G (1991) J Non-Crystal Solids 131–133:823
6. Kanetakis J, Fytas G, Hadjichristidis N (1991) Macromolecules 24:1806
7. Gerharz B, Fischer EW, Fytas G (1991) Polym Commun (in press)
8. Kanetakis J, Fytas G, Kremer F, Pakula T (1991) Macromolecules (submitted)
9. Stuhn B, Rennie AR (1989) Macromolecules 22:2460
10. Roland CM, Ngai KL (1991) Macromolecules 24:5315; and in this volume
11. Rizos AK, Ngai KL, Roland CM, Fytas G (in preparation)
12. Fytas G, Alig I, Rizos AK, Kremer F, Roovers JEL (in press)

Authors' address:

A. K. Rizos
Foundation for Research
and Technology Hellas
P.O. Box 1527
71 110 Heraklion
Crete, Greece

Progress in Colloid & Polymer Science

Progr Colloid Polym Sci 91:75–79 (1993)

Concentration fluctuations and segmental relaxation in miscible polymer blends

C. M. Roland and K. L. Ngai

Naval Research Laboratory, Washington, DC, USA

Abstract: Mechanical measurements in the glass transition zone were carried out on a series of polydienes to study the effect of chemical structure on the segmental relaxation. The breadth of the glass transition dispersion in the mechanical spectrum increased with increasing concentration of pendant vinyl groups on the main chain. The broader relaxations exhibited stronger temperature dependences. Both observations are consistent with the coupling model of relaxation. Enhanced intermolecular coupling arises due to steric interferences among the vinyl moieties, and the coupling model predicts that more cooperative relaxations will be both broader and more temperature sensitive. In miscible blends the glass transition dispersion is further broadened due the distribution of local environments engendered by concentration fluctuations. The coupling model is extended to consider the effect of concentration fluctuations on the local variation in the degree of intermolecular cooperativity. The shape of the relaxation spectra of miscible blends in the glass transition zone, as well as both composition and temperature dependencies of the dispersion, are markedly influenced by the concentration fluctuations specific to miscible blends. Application of the model to two blends is demonstrated to successfully describe the prominent features of the measurements.

Key words: Miscible blend – segmental relaxation – concentration fluctuations

Introduction

Although polymer mixtures exhibit a single glass transition similarly to neat polymers, their relaxation behavior is quite distinct. Relaxation of macroscopic variables in pure polymers near their glass transition almost invariably proceeds in accord with the KWW function

$$E(t) = (E_g - E_p) \exp\left[-\left(\frac{t}{\tau^*(n)} \right)^{1-n} \right] + E_p \, ,$$

$$(1)$$

in which E_g and E_p represent the glassy and relaxed moduli. This form for the relaxation function can be obtained by introducing intermolecular cooperativity, into expressions derived to describe relaxation of isolated chains [1, 2]. For the glass transition relaxation, the fundamental relaxation mode corresponds to intramolecularly correlated conformational transitions [3, 4]. However, as a segment tries to relax in dense phase, intermolecular couplings among the segments cause some of the attempted conformational transitions to fail. As dictated by cooperativity stochastic variations arise in the success rate for conformational transitions by the segments. When averaged over all segments this effect can be viewed as a slowing down (reduction) of the transition rate. The averaging appropriate for consideration of the relaxation of macroscopic variables leads to an effective description of the dynamics in terms of a time-dependent relaxation rate. Because of cooperativity the motions of individual segments are not identical at any given time, but the language of a time-dependent relaxation rate is retained in attempting to effectively describe the relaxation of macroscopic variables. On the molecular level there exists a "distribution of relaxation times", heterogeneous on the time scale of the effective relaxation time τ^*, as dictated by cooperativity in the success rate for conformational transitions. This underlying microscopic picture is consistent with recent experimental findings on segmental relaxation in poly(vinyl acetate) [5].

The τ^* in Eq. (1) depends on the strength of the intermolecular coupling as measured by the coupling parameter n [1]

$$\tau^*(n) = [(1 - n)\omega_c^n \tau^0]^{1/(1-n)}, \qquad (2)$$

where τ^0 is the relaxation time sans intermolecular coupling and ω_c^{-1} defines a characteristic time for the intermolecular couplings to become manifest (for neat polymers ω_c is typically of the order of 10^{10}–10^{11} sec^{-1}).

Pure polymers

Mechanical measurements were carried out in the vicinity of the glass transition for a series of polybutadienes of varying vinyl content (the 1,2-addition product) and for 1,4-polyisoprene [6]. The glass transition dispersion in the loss modulus could be well described by Eq. (1). The magnitude of the coupling parameter is observed to increase with increasing concentration of 1,2-chain units in the polymer backbone (Fig. 1). This result suggests that inflexible vinyl moieties projecting from the main chain effect interchain coupling via steric interactions. Enhanced intermolecular cooperativity results with relaxation of a segment more involves neighboring segments. It is particularly noteworthy that the most heterogeneous chain structure, corresponding to random copolymers of 1,2- and 1,4-units, have a narrower dispersion (smaller n) than the pure homopolymer 1,2-polybutadiene. This result is difficult to reconcile with a simple distribution of relaxation times as the origin of the spectral broadening. Any model that attempts to describe segmental relaxation in bulk polymers solely in terms of intrachain conformational transition rates will be unable to account for the data in Fig. 1.

The coupling model predicts that there is a correlation between time and temperature dependencies of the relaxation [7]. Specifically, polymer chains whose segmental relaxation is characterized by stronger intermolecular coupling (larger n) should exhibit a more marked dependence on tem-

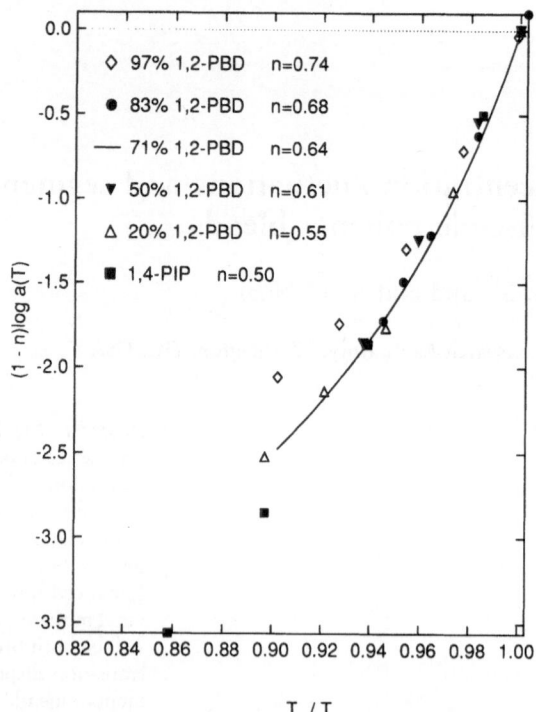

Fig. 1. The temperature dependence of the shift factors for the various polydienes, with data from ref. [8] also included. The inverse temperature has been normalized by T_g^{-1}, with the glass transition temperatures taken to be the temperature at which the apparent relaxation time equaled 100 s

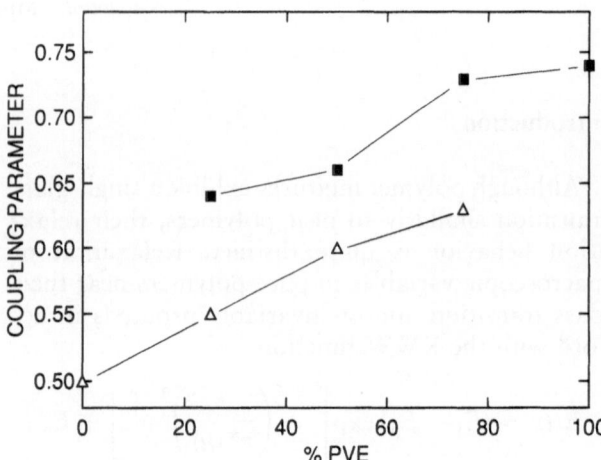

Fig. 2. The mean coupling parameter measured for the components as a function of the blend composition

perature. As demonstrated in Fig. 1 this correlation of the frequency and temperature dependencies is observed for the polydiene series [6], including values for a 71% 1,2-polybutadiene taken from the literature [8]. Note that the ordinate scale includes the factor $(1 - n)$, which causes the data to superimpose.

Blends

Unlike pure polymers, the segmental relaxation behavior of miscible blends usually does not conform to Eq. (1). The mechanical and dielectric glass transition dispersions are broader, often exhibiting an extraordinary low-frequency tail, along with a strong temperature dependence. The origin of this broadening is the distribution of segment environments engendered by concentration fluctuations. These concentration fluctuations are intrinsic to thermodynamically miscible multicomponent compositions and are governed by the free energy of mixing and its composition dependence. However, at least well away from the two-phase region of the phase diagram, the length scale of the concentration fluctuations may be such that, together with a weak temperature dependence, the consequent spectral broadening is not directly dependent on thermodynamic considerations. A model has been proposed to account for the composition dependence and shape of the relaxation spectra of miscible blends in the glass transition zone. A single blend of fixed composition possesses a myriad of local environments due to concentration fluctuations. This distribution of local environments effects a distribution of both the τ^0's and the coupling parameters, i.e., concentration fluctuations affect the local friction coefficient as well as the degree to which a given segment's relaxation involves neighboring chain units. The measurement of a macroscopic variable such as mechanical stress or dielectric response reflects the individual contributions arising from the distinct local environment. The interaction of these environments (e.g., local field effects in dielectric spectroscopy or homogeneity of strain or stress in mechanical measurements) will influence the manner in which the various local contributions are manifested in the macroscopic measurement. If uniformity of local strain (no local field corrections) is assumed, the mechanical or dielectric loss spectra measured isothermally can be described using the expression

$$\varepsilon''(\omega) = \operatorname{Im} \sum_j \int_0^\infty \Delta\varepsilon_j \int_{l_l}^{l_u} -\frac{d}{dt}(1/N_j)$$

$$\times \left\{ \exp\left[-a_j(n - n_j)^2\right] \right.$$

$$\times \exp\left[-\left(\frac{t}{\tau^*(\tau_j^0, n)}\right)^{1-n}\right]\right\}$$

$$\times \exp\left[-i\omega t\right] dt\, dn , \tag{3}$$

where $\Delta\varepsilon_j$ is the relaxation strength of the jth component, n_j is the mean, a_j the variance and N_j the normalization constant of the Gaussian distribution of coupling parameters, and $\tau^*(\tau_j^0, n)$ is calculated from Eq. (2). In principle the integration limits range from $n = 0$ (corresponding to no intermolecular cooperativity) through $n = 1$ (cessation of relaxation). In reality the range of n is determined by the extremes in local composition and the corresponding degrees of intermolecular coupling. On the right-hand side of Eq. (3), for each value of n in the normal distribution of the jth component, $\tau^*(\tau_j^0, n)$ is calculated according to Eq. (2) and the correlation function, $\exp[-(t/\tau^*(\tau_j^0, n))^{1-n}]$, can be written down. Taking the Fourier transform of the time derivative of this correlation function, multiplying by the relaxation strength $\Delta\varepsilon_j$ of the jth component and taking the imaginary part gives the contribution to the dielectric (or mechanical) loss. Integrating this contribution over the normal distribution of n (from the distribution in local environments) and summing over the j components finally yields the total response as a function of frequency.

For the sake of simplicity, a distribution of τ^0's, that may or may not be correlated with the distribution of n's, is not included in Eq. (3). For a detailed analysis of relaxation data from blends it may prove necessary to introduce such a distribution of τ^0's in future work. For present purposes we have only included in Eq. (3) the effects that a distribution of coupling parameters would have on the blend dynamics. This aspect of blends is focussed on herein because such a distribution of coupling parameters does not occur in pure (unblended) polymers. It is a novel feature of miscible blends. While a distribution of τ^0's is likely present in blends given the distribution of local environments, an analysis of the combined effect of distributions in τ^0 and n is not readily carried out with the available data. Each τ^0 will have its own set of Vogel–Fulcher parameters to describe its temperature dependence. The profusion of factors necessary to be taken into account makes a unique theoretical description problematical at the present time. The coupling model approach to blend dynamics is necessary, however, in light of its singular ability to explain many properties observed in neat polymer dynamics.

Equation (3) was applied to mechanical measurements on blends of 1,2-polybutadiene (PVE) and 1,4-polyisoprene (PIP), the polydienes in Fig. 1 exhibiting the most diverse segmental relaxation behavior. These polymers form thermodynamically miscible blends without any specific interactions [9, 10]. At high concentrations of PVE the breadth of the loss modulus peak associated with the glass transition in the blend is anomalously broad, extending over many decades of frequency [11, 12]. The application of Eq. (3) to the experimentally measured mechanical loss spectra for three blend compositions demonstrate that the salient features of the experimental data can be explained rather well with this approach. A PVE-rich environment is more effective in coupling to the primitive relaxation ($n = 0.74$ and $n = 0.50$ in PVE and PIP, respectively from Fig. 2). Additionally, since the glass transition temperature of PVE is 70° higher than that of PIP, PIP segments surrounded by PVE chain units will be relaxing in an environment that itself is relatively unrelaxed and, thus, unaccommodating. For these reasons local blend environments richer in the PVE are expected to confer to locally rearranging segments a larger degree of intermolecular coupling. This effect underlies the characteristic broadening on the low-frequency side of the glass transition dispersion. Note that if the effect of concentration fluctuations were only to displace the contributions from the segments to various positions about the peak frequency, the spectral broadening would be symmetric. It is the distribution in intermolecular cooperativity engendered by the fluctuations that gives rise to the observed asymmetry.

Dielectric relaxation studies of blends can offer advantages over mechanical measurements since in favorable cases one component will be the main contributor to the dielectric spectrum. Since poly(vinyl methyl ether) (PVME) has a much larger dipole moment than does polystyrene (PS), it dominates the dielectric response of their blends. From an analysis of dielectric measurements, one can determine the effect of changing the local environment on segmental relaxation of PVME. By fitting Eq. (3) to isothermal data obtained on a mixture containing 40% PS, the dependence of n for PVME on local composition was deduced [13]. The higher frequency side of the dispersion reflects the PVME-rich environs, and the lower frequency response arises from PVME in PS-rich regions. The coupling parameter associated with local environments rich in PVME is found to be close to that of pure PVME ($n = 0.56$). An increasing local concentration of PS has the effect of increasing the degree of intermolecular coupling, with a value of n as large as 0.75 deduced for PVME contributing in the low-frequency tail of the spectrum.

It is found once again, as predicted by the coupling model [7], that a correlation exists between the magnitude of the intermolecular coupling and the temperature dependence of the relaxation time. As shown in Fig. 3, the shift factors for PVME segments in local environments characterized by larger n change more with temperature than segments whose relaxation is less cooperative [13].

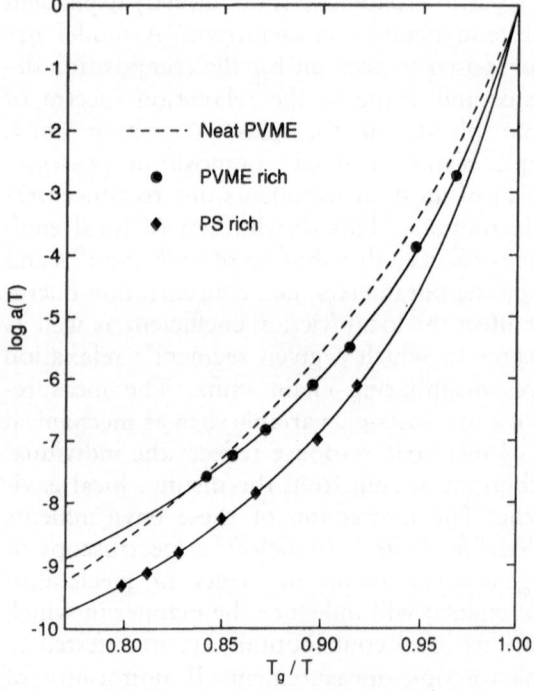

Fig. 3. The inverse frequency, normalized by its value at the blend glass transition temperature ($T_g = 265$ K), as a function of the inverse temperature normalized by T_g^{-1} for pure PVME (−−−) and for the blend with 40% PS (——). Each curve for the blend corresponds to a particular part of the spectral dispersion, corresponding to the local composition whose coupling parameter is indicated. The contribution on the high-frequency side is seen to approach that of the pure PVME

References

1. Ngai KL, Rendell RW, Rajagopal AK, Teitler S (1986) Ann NY Acad Sci 484:150
2. Ngai KL, Rendell RW (1991) J Non-Cryst Solids 131–133:942
3. Hall CK, Helfand E (1982) J Chem Phys 77:3275
4. Bahar I, Erman B, Monnerie L (1991) Macromolecules 24:3618
5. Schmidt-Rohr K, Spiess HW (1991) Phys Rev Lett 66:3020
6. Roland CM, Ngai KL (1991) Macromolecules 24:5315
7. Plazek DJ, Ngai KL (1991) Macromolecules 24:1222
8. Floudas G, PhD Thesis (1990) University of Crete, Herkalion, Crete
9. Roland CM (1987) Macromolecules 20:2557
10. Roland CM (1988) J Poly Sci Poly Phys Ed 26:839
11. Trask CA, Roland CM (1989) Macromolecules 22:256
12. Roland CM, Ngai KL (1991) Macromolecules 24:2261
13. Roland CM, Ngai KL (1991) Macromolecules, in press

Received November 13, 1991;
accepted June 6, 1992

Authors' address:

C. M. Roland
Naval Research Laboratory
Code 6120
Washington, DC, 20375-5000, USA

Progress in Colloid & Polymer Science

Progr Colloid Polym Sci 91:80–82 (1993)

Exponential and non-exponential relaxation and early state of spinodal decomposition in polymer blends by SANS

D. Schwahn, S. Janßen, and T. Springer

Forschungszentrum Jülich GmbH, Institut für Festkörperforschung, Jülich, FRG

Abstract: With neutron small-angle scattering it was found, for two different polymer blends, that the time-dependent structure factor $S(Q, t)$ relaxes exponentially or also non-exponentially after a temperature step has been applied. The classical Cahn–Hilliard–Cook theory predicts an exponential time behaviour of $S(Q, t)$. The observed deviations may be due to the correlation of the order parameter with internal motions of the polymer segments or with other internal variables of a viscoelastic material near its glass transition.

Key words: Polymer blends – spinodal decomposition – relaxation – segmental motion – viscoelastic material

We discuss the evolution of the time-dependent structure factor $S(Q, t)$ from small-angle neutron scattering (SANS) as concerns the relaxation and the early state of spinodal decomposition in polymer blends (Q is the scattering vector). The classical Cahn–Hilliard–Cook (CHC) theory predicts an exponential time dependence of the structure factor, namely [1, 2]

$$S(Q, t) = S_T(Q)$$
$$+ [S_{T_0}(Q) - S_T(Q)] \, L^2(Q, t) , \quad (1)$$

with $L(Q, t) = \exp[R(Q) t]$. The corresponding relaxation rate is $R(Q) = Q^2 \Lambda(Q) S_T^{-1}(Q)$, where $\Lambda(Q)$ is the non-local Onsager coefficient [3, 4]. Equation (1) describes the transition from the equilibrium state at the initial temperature T_0 to the equilibrium state at T, after a temperature step $T_0 \rightarrow T$ has been imposed onto the blend. The two states are characterized by the static equilibrium structure factors $S_{T_0}(Q)$ and $S_T(Q)$, respectively. In the case that T is in the *unstable* region, $S_T(Q)$ is obtained following the procedure as described in detail by Cook [2], namely by an extrapolation of the experimental data for sufficiently large Q values and for times when coarsening does not yet come into play, i.e. as long as the (linear) CHC theory holds. This should then lead to a function

$S_T(Q)$ which consistently describes *all* data by Eq. (1).

The volume fraction of one of the components ϕ is the order parameter and the only variable in the process leading to Eq. (1), which means that for polymers the fast "segmental" or internal modes should have died out. However, internal modes are observable in the early state of a transition from an equilibrium state only up to a time [5]:

$$\tau_R = \langle R^2 \rangle / 2D_{tr} = \langle R^2 \rangle / 2\Lambda(Q = 0)\chi_s , \quad (2)$$

where $\langle R^2 \rangle$ is the mean square of the end-to-end vector of the polymer, χ_s is the Flory–Huggins interaction parameter of the blend at the spinodal, and D_{tr} is the tracer diffusion constant. τ_R is the time the polymer needs to diffuse over a distance equal to its own size.

Figure 1a and b shows the time-dependent structure factor from SANS experiments $S(Q, t)$ for the blend: d-PS/PVME, with

$$V_w(\text{d-PS}) = 0.34 \cdot 10^6 \text{ cm}^3/\text{mol} \quad \text{and}$$

$$V_w(\text{PVME}) = 0.63 \cdot 10^5 \text{ cm}^3/\text{mol} .$$

The blend was at the critical composition ϕ(d-PS) = 0.13. It has a lower critical solution temperature at $T_c = 132.08\,°C$ as determined by critical

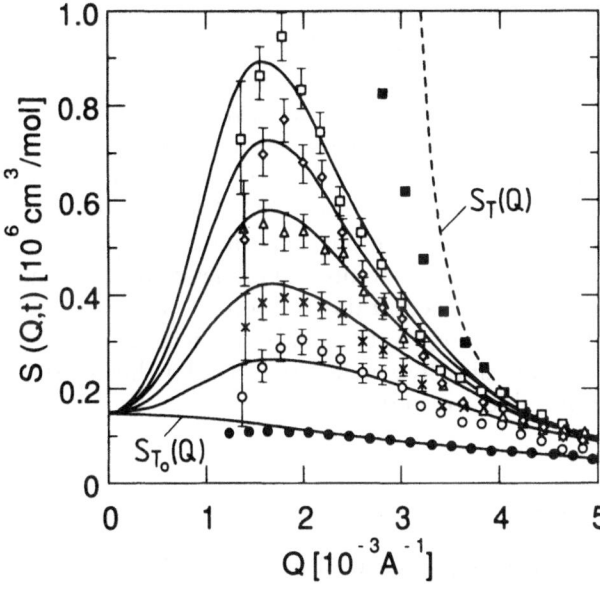

Fig. 1a. The time-dependent structure factor from SANS experiments $S(Q, t)$ for the early states of spinodal decomposition of the system d-PS/PVME, after a temperature step from $T_0 = 129.85\,°C$–$134.1\,°C$ has been applied. The critical temperature is $T_c = 132.08\,°C$. The period of one measurement is 10 s, carried out after $(-\bigcirc-)$ 65 s, $(-\times-)$ 115 s, $(-\triangle-)$ 155 s, $(-\diamondsuit-)$ 185 s, $(-\square-)$ 215 s, and $(-\blacksquare-)$ 805 s. The $(-\blacksquare-)$ data represent $S_T(Q)$ for $Q \gtrsim 3.7 \cdot 10^{-3}\,\text{Å}^{-1}$

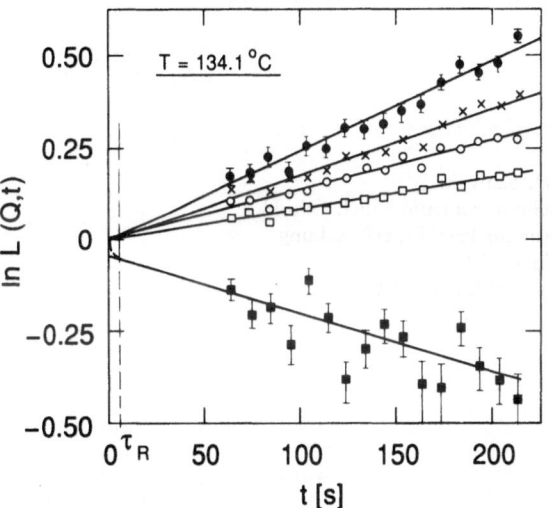

Fig. 1b. Semi-logarithmic presentation of the relaxation part $\ln L(Q, t)$ according to Eq. (1), as a function of time for different Q, namely $(-\bullet-)$ $Q = 1.79$, $(-\times-)$ 2.19, $(-\bigcirc-)$ 2.4, $(-\square-)$ 2.6, and $(-\blacksquare-)$ 3.62 (in units of $10^{-3}\,\text{Å}^{-1}$). The straight lines indicate an exponential time behaviour according to the CHC theory. The relaxation frequency $R(Q)$ is obtained from the slope of these lines

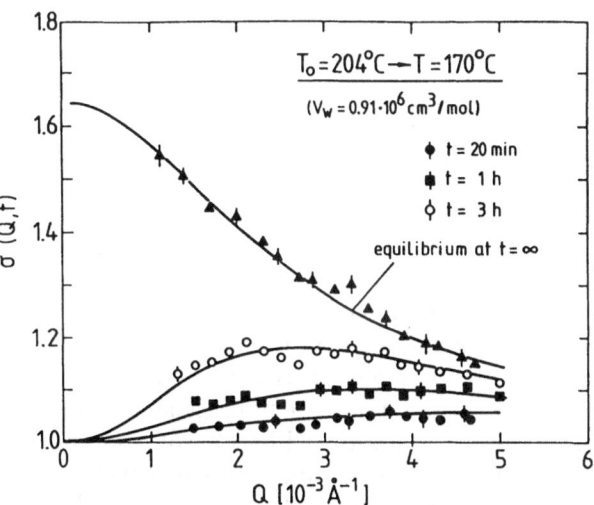

Fig. 2a. Reduced time-dependent structure factor $\sigma(Q, t) = S_T(Q, t)/S_{T_0}(Q)$, where S_{T_0} is the initial equilibrium state of the structure factor

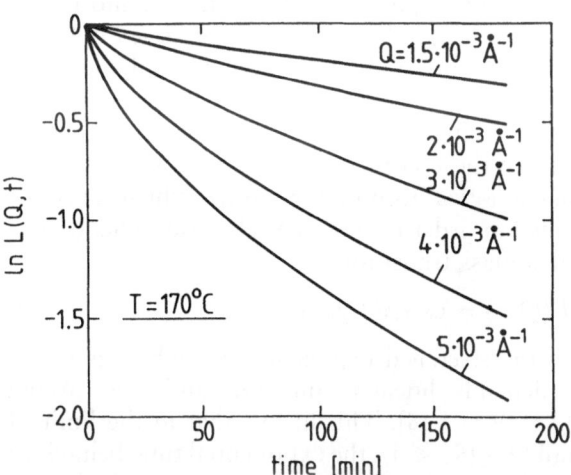

Fig. 2b. Logarithm of the relaxation function as a function of time. The deviation from an exponential time-behaviour is stronger at larger Q values. In the limit of $Q = 0$, $L(Q, t)$ becomes a single exponential

scattering [6]. The transition started at the equilibrium state $S_{T_0}(Q)$ two degrees below T_c, and led to a value of two degrees above T_c in the unstable region. The measurements were performed in "real time" and the time used for each scattering curve was 10 s. The interference peak typical for spinodal decomposition appears at $1.5 \cdot 10^{-3}\,\text{Å}^{-1}$. The data measured up to 215 s could be interpreted by the linear CHC theory, where the extrapolated structure factor $S_T(Q)$ is plotted in Fig. 1a as a dotted line. $L(Q, t)$ is plotted in Fig. 1b for various Q values as a function of time t in a semi-logarithmic scale. The least-squares fit shows a linear t-dependence for all Q values according to a single exponential behaviour of the CHC theory. However, in Fig. 1b a non-exponential behaviour appears at the largest $Q = 3.62 \cdot 10^{-3} \, \text{Å}^{-1}$ for $t < \tau_R$, where $\tau_R = 13\,\text{s}$ following Eq. (2). A detailed discussion of these experiments is the subject of a forthcoming paper [7].

Figure 2a and b shows the SANS results of a relaxation experiment between two equilibrium states in the one-phase or homogeneous region of the isotope mixture d-PS/PS of polystyrene ($V_w = 0.91 \cdot 10^6 \, \text{cm}^3/\text{mol}$, $\phi(\text{d-PS}) = 0.48$, $T_c = 127\,°\text{C}$; see [8]). Figure 2b presents again $L(Q, t)$ versus time for various Q values which were obtained from a fit to the experimental data in Fig. 2a (solid lines). A clear deviation from an exponential time law is observed. The time constant τ_R is estimated from Eq. (2) to be 7.9 h, which is more than two times larger than the annealing time t in Fig. 2. Therefore, internal modes of the polymer may come into play, being responsible for the deviation from the single exponential behaviour. We have analysed these non-exponential data with the phenomenological Kohlrausch ansatz, which is usually applied in order to describe relaxation phenomena near a glass transition:

$$L(Q, t) = \exp\left[R(Q) \, t \right]^{\alpha(Q)} \tag{3}$$

with the stretched exponent $\alpha(Q)$ which is found to depend linearly on Q, and we found $\alpha(Q = 0) = 1$ [8]. This means that in the limit of small Q ($QR_g \ll 1$), the exponential time behaviour according to the CHC theory is approached.

There are two explanations for non-exponential behaviour: The PS has a glass transition at $T_G = 100\,°\text{C}$. Under these conditions, the order parameter may be correlated with the internal variables of a viscoelastic material [9]. Or, on the other hand, the internal motions of the polymer segment are dominant [10]. This may lead to two or more exponentials which fit our phenomenological stretched-exponent formula. This is consistent with the fact that our experimental time is still below τ_R, such that internal modes could come into play. On the contrary, for modes with very long wavelength (Q very small) the relaxation is expectedly diffusive such that $\alpha = 1$. Further work on this subject is under way.

References

1. Cahn JW (1961) Acta Metallogr 9:795–801
2. Cook HE (1970) Acta Metallogr 18:297–306
3. de Gennes PG (1979) Scaling Concepts in Polymer Physics. Cornell University Press, Ithaca
4. Binder K (1983) J Chem Phys 79:6387–6409
5. Strobl GR (1985) Macromolecules 18:558–563
6. Schwahn D, Springer T, Janßen S (1991) Physica B 174:159–163
7. Schwahn D, Janßen S, Springer T (1992) J Chem Phys, accepted for publication
8. Schwahn D, Hahn K, Streib J, Springer T (1990) J Chem Phys 93:8383–8391
9. Jäckle J, Pieroth M (1988) Z Phys B 72:25–39
10. Akcasu AZ (1989) Macromolecules 22:3682–3689

Received January 23, 1992;
accepted May 28, 1992

Authors' address:

Dr. D. Schwahn
Forschungszentrum Jülich GmbH
Institut für Festkörperforschung
Postfach 19 13
D-W-5170 Jülich, FRG

Progress in Colloid & Polymer Science   Progr Colloid Polym Sci 91:83–87 (1993)

Equilibrium and dynamic aspects of end-attached diblock and triblock copolymer chains

C. Toprakcioglu[1,2]), L. Dai[1]), M. A. Ansarifar[1,*]), M. Stamm[3]), and H. Motschmann[3])

[1]) Cavendish Laboratory, University of Cambridge, England
[2]) Institute of Food Research, AFRC, Norwich, England
[3]) Max-Planck Institute fur Polymerforschung, Mainz, FRG

Abstract: We report results on adsorbed layers of polystyrene (PS)–polyethylene oxide (PEO) diblock and PEO–PS–PEO triblock copolymers at the solid/liquid interface. The volume fraction profiles of the adsorbed PS–PEO diblock copolymers at the quartz/toluene interface determined by neutron reflectometry are consistent with a parabolic profile as predicted by theory. Ellipsometry was used to investigate the adsorption kinetics of PS–PEO diblock copolymer chains on SiO_2 from toluene. The results show a rapid diffusion-controlled process, characterized by a linear increase in the adsorbed amount with the square-root of time, at the early stages of adsorption, while at late times adsorption is slow and appears to be governed by an exponential time behavior. Studies of the PEO–PS–PEO copolymer, using the surface-force apparatus, reveal conformational rearrangements for the adsorbed triblock chains between mica surfaces. When two such surfaces are in close proximity to each other, triblock chains originally adsorbed in a "loop" conformation may open into a "bridge" conformation with both PEO blocks of a given chain adsorbing onto opposing mica surfaces. This is indicated by the appearance of clearly detectable attractive forces and their evolution with compression–decompression cycles. The time evolution of these forces reflects the dynamics of the rearrangement of chains between the two surfaces.

Key words: Adsorption – block copolymers – ellipsometry – neutron reflectometry – surface-force apparatus

Introduction

In general, the conformation of polymer chains adsorbed at the solid/liquid interface differs significantly from that in bulk solution. When a solid surface comes into contact with a suitable macromolecular solution, polymer chains may adsorb readily on the solid/liquid interface leading to the formation of a layer whose polymer concentration or volume fraction is much greater than the corresponding value in the bulk solution. As the polymer concentration in the adsorbed layer increases with the adsorption process, interactions between the adsorbed chains are enhanced and their average conformation changes continuously towards an equilibrium state which is generally very different from the initial one. This conformational variation is particularly dramatic in the case of end-adsorbed or "tethered" chains in a good solvent, where random coils of the polymer chains in solution become strongly stretched polymer "brushes" at the interface after a prolonged process of adsorption [1].

The kinetics of adsorption may be expected to reflect such conformational changes. At equilibrium, the polymer density profile normal to the interface reflects the final conformation of the

---

*) Current Address: The Malaysian Rubber Producers Association, Brickendonburry, Hertfordshire, SG13 8NL, England

chains in the adsorbed layer. The forces between such an adsorbed layer and an opposing solid wall provide further insight into the conformation of the end-adsorbed chains. In the present paper, we present ellipsometry, neutron reflectivity, and mica force balance data to address these issues. The fact that PEO segments can readily adsorb on mica or quartz, whereas PS does not adsorb from toluene make the PS–PEO diblock and PEO–PS–PEO triblock copolymers the systems of choice for such a study. A number of previous investigations on the PS–PEO diblock copolymers, particularly using the surface-force apparatus [1–3], provide important clues for interpreting the force data from the PEO–PS–PEO triblock chains.

Materials and methods

The PS–PEO and PEO–PS–PEO copolymer samples were purchased from Polymer Laboratories (UK) and have the molecular characteristics shown in Table 1. Spectroscopic grade toluene and deuterated toluene (d-toluene) obtained from Aldrich Chemical Company were used as supplied.

For the neutron reflectivity measurements an optically flat single crystal quartz surface was used as a substrate, while for ellipsometry, mirror-polished silicon wafers with a well-defined SiO_2 layer obtained from Wacker Chemie were used for this purpose. Both the ellipsometric technique and the neutron reflectivity method are described in detail elsewhere [4, 5]. The procedures for using the surface-forces apparatus to measure the force–distance profiles of *single* adsorbed polymer layers against a *bare* mica surface have been recently reported [6].

Results and discussion

PS–PEO diblock copolymers

Ellipsometry was used to study the adsorption kinetics of the PS–PEO copolymers onto a SiO_2/toluene interface by measuring the total adsorbed amount of polymer, A, as a function of time, t. It may be reasonably assumed that at the initial stages of adsorption PS–PEO chains can adsorb readily onto the substrate upon reaching the solid/liquid interface, and that this process goes on until the polymer concentration in the adsorbed layer approaches a value above which the end-attached non-adsorbing PS chains begin to interact. Such a diffusion-controlled process is clearly evidenced by the well-defined linear dependence of A on $t^{1/2}$ at early times, as shown in Fig. 1. The diffusion constants obtained from the gradients of these plots are in good agreement with those reported for PS chains under similar conditions [4]. As the adsorption proceeds further, however, more PS–PEO chains adsorb at the interface and the end-attached PS chains experience an increasing osmotic repulsion thereby becoming strongly stretched away from the interface. Thus, the adsorption of addition chains, beyond this stage, is hindered by the barrier due to the chains already adsorbed on the substrate and involves significant conformational rearrangement both for attached and in-coming chains. Consequently, the adsorption process slows down, as indicated by the negative deviation from the linear behavior of A vs. $t^{1/2}$ shown in Fig. 1. The increase of adsorbance with t at these later stages appears to follow an exponential function. The final attainment of equilibrium is, thus, a slow process and the adsorption kinetics is closely related to the conformational changes of

Table 1. Molecular characteristics of the PS–PEO and PEO–PS–PEO copolymer samples

Sample	M_w	M_w/M_n	wt% PEO	$(PEO)_{x1}$ x1	$(PS)_y$ y	$(PEO)_{x2}$ x2
PS–PEO (80 k)	$80 \cdot 10^3$	1.07	5.0	0	730	90
PS–PEO (150 k)	$150 \cdot 10^3$	1.16	1.5	0	1420	51
PS–PEO (184 k)	$184 \cdot 10^3$	1.10	4.0	0	1700	167
PS–PEO (502 k)	$502 \cdot 10^3$	1.10	0.8	0	4788	91
PEO–PS–PEO (128 k)	$128 \cdot 10^3$	1.02	0.3	5	1225	5

x1, x2 and y are weight-averaged degrees of polymerization for each block.

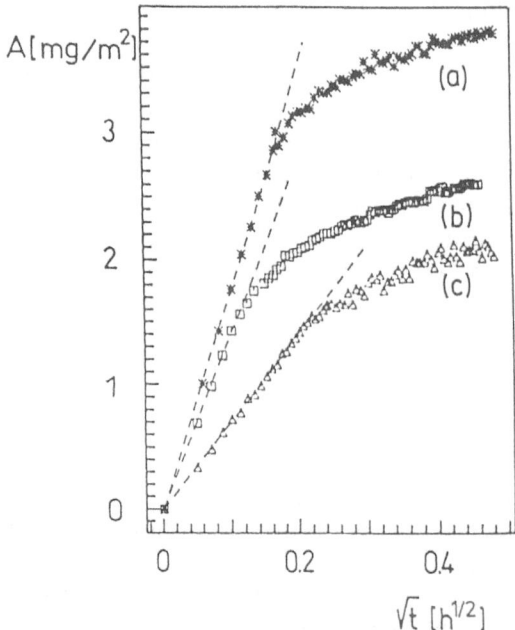

Fig. 1. Plot of adsorbed amount of diblock copolymer, *A*, vs. adsorption time, *t*, at polymer solution concentrations $C = 0.015$ (mg/ml). a) PS–PEO (80 K); (b) PS–PEO (184 K); (c) PS–PEO (502 K). The substrate is SiO_2, and the solvent is toluene

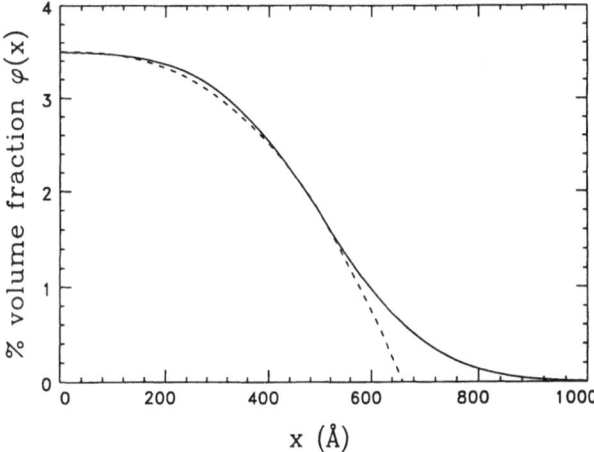

Fig. 2. Error function (——) and parabolic (– – – –) polymer density profiles for PS–PEO (150 K) adsorbed on quartz from toluene, determined by neutron reflectometry. The curves were obtained from least-squares fits of each model to the experimental reflectivity data (not shown). The two profiles are essentially indistinguishable at short to intermediate distances from the surface, and differ only at large distances near the "height" of the polymer brush

the adsorbed chains as a function of local polymer concentration.

The equilibrium polymer density profile in an end-adsorbed PS–PEO layer at the quartz/*d*-toluene interface was investigated by neutron reflectometry [5]. The results are well-described by a density profile of largely parabolic shape with a Gaussian-like "tail" (Fig. 2), which is in good agreement with the theoretical prediction for semidilute polymer brushes [7].

PEO–PS–PEO triblock copolymers

While AB-type diblock copolymers containing short "sticking" segments of A, and non-adsorbing B blocks can only adsorb onto a substrate in a "tail" conformation, ABA triblock chains can also form "loops" with both A blocks in a single chain adsorbing on the same substrate. Under certain conditions, moreover, ABA triblock copolymers may form polymer "bridges" by simultaneously attaching both of the A blocks in the same chain onto two opposing substrates. We have studied these conformational features by measuring the interactions between *single* layers of end-adsorbed PEO–PS–PEO triblock copolymers and a *bare* mica surface.

To begin with, the interactions of two bare mica surfaces in toluene and a single adsorbed PS–PEO layer against a bare mica surface were measured as a control. The force–distance profile seen in Fig. 3a is very similar to those previously reported for the bare mica surfaces in undried toluene [1, 6, 8]. Since only a tail conformation is possible for a PS–PEO diblock copolymer adsorbed on mica from toluene, the interaction of such a PS "brush" with a bare mica surface should be monotonically repulsive, as is observed in Fig. 3b where attractive forces are completely absent. As mentioned above, however, for PEO–PS–PEO triblock copolymers both loops and tails may be formed at the mica/toluene interface. The fraction of tail chains in the adsorbed layer is a function of the adsorption energy of the "sticking" block (PEO) and could be negligibly low if the interaction energy between the PEO block and the surface is sufficiently high, which normally requires rather long PEO blocks in the macromolecule. It is worth noting that the PEO–PS–PEO triblock copolymer used in this study contains very short PEO blocks (see Table 1), and that a significant number of tails may, thus, be

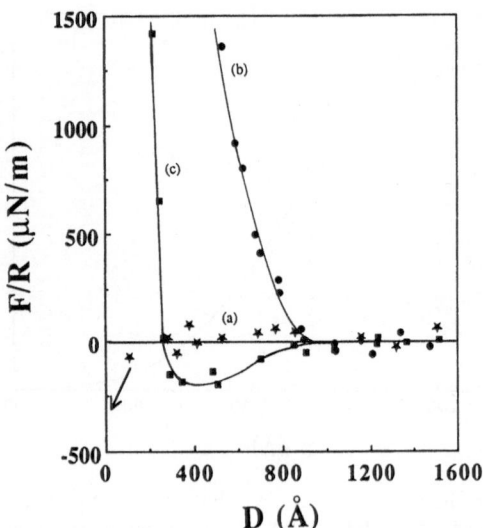

Fig. 3. Force–distance profiles: (a) between two curved bare mica surfaces in undried toluene; (b) a single PS–PEO (150 K) adsorbed layer against a bare mica surface in toluene; (c) a single PEO–PS–PEO (128 K) adsorbed layer against a bare mica surface in toluene at the sixth compression–decompression cycle. The radius of curvature of the mica surfaces is $R \approx 10$ mm in each case

expected in the adsorbed layer. This is confirmed by the force–distance profile for the interaction of a single PEO–PS–PEO adsorbed layer with a bare mica surface (Fig. 3c). An attractive profile is observed, in contrast to the monotonic repulsion associated with the diblock copolymer (Fig. 3b). The attraction seen in Fig. 3c unambiguously indicates the presence of tails in the adsorbed triblock copolymer layer, as the free PEO end-blocks of chains in tail conformation can readily adsorb onto the bare mica sheet forming polymer "bridges" between the surfaces, which leads to an attraction between the adsorbed layer and the bare mica sheet. It should also be noted that the steep repulsive wall in the force–distance profile of the end-adsorbed PEO–PS–PEO triblock layer (Fig. 3c) occurs at a much shorter separation of the surfaces than the corresponding case of the PS–PEO diblock layer (Fig. 3b), even though these materials have comparable molecular weights. This presumably indicates that while the adsorbed PEO–PS–PEO layer contains some fairly stretched "tail" chains, there is also a very substantial number of chains bound in a "loop" conformation which reduce significantly the interaction range of repulsion.

Furthermore, it is also found that repeated compression-decompression cycles cause further con-

formational changes resulting in enhancement of the attractive forces over time (not shown), and the eventual distribution of the adsorbed polymer chains equally between the two opposing surfaces [6]. The dynamics of these chain rearrangements are indicated by a time evolution of the force–distance profiles typified by a change from attraction to monotonic repulsion. These results have potentially important implications for the use of block copolymers as stabilizers and/or flocculants for colloidal dispersions.

Conclusions

We have demonstrated that the formation of an end-attached polymer brush at a solid/liquid interface in a good solvent is characterized by adsorption kinetics with two distinct regimes: a rapid diffusion-controlled process at initial stages, followed by a much slower regime at later times and an exponential approach to equilibrium. At equilibrium, the adsorbed polymer density profile normal to the interface is found to be well-described by a parabolic function as predicted by theory. Furthermore, force measurements between a single end-adsorbed polymer layer and a bare mica surface are sensitive to the conformation of the end-attached chains at the solid/liquid interface. Particularly, it is found that in contrast to the end-adsorbed PS–PEO diblock copolymers which show only a repulsive interaction with a bare mica surface in toluene, the end-adsorbed PEO–PS–PEO triblock copolymers exhibit attractive interactions with the bare mica surface due to the formation of polymer bridges between the two mica surfaces in the good solvent by simultaneously anchoring each of the two PEO blocks in a given chain onto opposing mica sheets.

References

1. Taunton HJ, Toprakcioglu C, Fetters LJ, Klein J (1990) Macromolecules 23:571–580; (1988) Nature 332:712–714
2. Taunton HJ, Toprakcioglu C, Klein J (1988) Macromolecules 21:3333–3336
3. Guzonas D, Boils D, Hair ML (1991) Macromolecules 24:3383–3387
4. Motschmann H, Stamm M, Toprakcioglu C (1991) Macromolecules 24:3681–3688
5. Field JB, Toprakcioglu C, Ball RC, Stanley HB, Dai L,

Barford W, Penfold J, Smith G, Hamilton W, Macro-molecules, in press
6. Dai L, Toprakcioglu C (1991) Europhys Lett 16:331–335
7. Milner ST, Witten TA, Cates MF (1988) Europhys Lett 5:413–418; (1988) Macromolecules 21:2610–2619
8. Luckham PF, Klein J (1985) Macromolecules 18:721–728

Received December 10, 1991;
accepted May 20, 1992

Authors' address:

Dr. Chris Toprakcioglu
Cavendish Laboratory, PCS
University of Cambridge
Madingley Road
Cambridge, CB3 0HE, England

Progress in Colloid & Polymer Science   Progr Colloid Polym Sci 91:88–92 (1993)

Very thin films of symmetric diblock copolymers

S. H. Anastasiadis*), A. Menelle, T. P. Russell, S. K. Satija[1]), and C. F. Majkrzak[1])

IBM Research Division, Almaden Research Center, San Jose, USA
[1]) National Institute of Standards and Technology, Reactor Radiation Division, Gaithersburg, USA

Abstract: The morphology of thin films of symmetric diblock copolymers has been investigated by neutron and X-ray reflectivity and X-ray photoelectron spectroscopy, for thicknesses less than $3L/2$, where L is the long period of the lamellar morphology in the bulk. The constrains placed on a copolymer film by the presence of the two surfaces can produce severe perturbations on the morphology, due to the interactions of the copolymer blocks with the air and the substrate interfaces. It is possible to induce variations in the chain extension and in the interfacial mixing, and to force a normally microphase separated copolymer into a homogeneous state.

Key words: Diblock copolymers – thin films – neutron reflectivity

Introduction

The morphology of symmetric diblock copolymers in the bulk is determined by the total number of copolymer segments, N, and the Flory–Huggins segmental interaction parameter, χ. It is well known that in the bulk and at temperatures such that $\chi N > 10.5$ symmetric diblock copolymers exhibit an ordered lamellar microdomain morphology, whereas for higher temperatures a disordered morphology exists. Thin films of diblock copolymers represent a situation where the copolymer experiences restraints induced by the proximity of the two interfaces, these being the substrate/polymer and the polymer/air interfaces. When the film thickness is several times the characteristic period of the copolymer in the bulk, the specific interactions of the blocks with either interface lead to an orientation of the lamellar microdomains absolutely parallel to the surface [1, 2]. The period of the multilayered morphology is shown to be that of the bulk copolymer. The presence of the interfaces can also induce ordering of the diblock copolymers even at temperatures well above the bulk order–disorder transition [3, 4], in qualitative agreement with mean field arguments [5].

Decreasing the thickness of the film such that it approaches or is less than one period, L, of the multilayered structure places further constraints on the copolymer molecules. It is the intent of this article to address the question of the effect of these surface constraints on the morphology of very thin films using neutron reflectometry on thin films with thickness less or equal to $3L/2$.

Experimental

The copolymer used in this study is a symmetric diblock copolymer of polystyrene, PS, and polymethylmethacrylate, PMMA, denoted P(D-S-b-MMA), where the styrene block is deuterated. The copolymer has 1513 PS and 1319 PMMA segments, which yields a fraction of PS segments of 0.53. The

---

*) Present adress: Foundation for Research and Technology-Hellas, Institute of Electronic Structure and Laser, P.O. Box 1527, 711 10 Heraklion Crete, Greece

polydispersity is 1.08. Thin films of the copolymer were prepared by spin-coating solutions of P(D-S-b-MMA) in toluene onto 10 cm in diameter, 5 mm thick polished Si 100 substrates. All films were annealed for 240 h at 170 °C to ensure that the films had achieved thermal equilibrium.

Neutron reflectivity measurements were performed on the BT-4 triple axis diffractometer at the Reactor Experimental Hall of the National Institute of Standards and Technology [2]. A collimated monochromatic neutron beam with a wavelength, λ, of 2.35 Å, $\Delta\lambda/\lambda = 0.02$ and an angular divergence of 0.02 ° illuminated the specimen at glancing angles. Reflectivity profiles were obtained with the specimen rotating at an angle θ and the detector at an angle 2θ, keeping the diffraction vector perpendicular to the sample surface. The background was measured with the detector offset by $+0.6$ ° from the specular position and performing the normal θ–2θ scan.

Results and discussion

Films with five different thicknesses were investigated in this study. In particular, films with approximate thicknesses of $L/4$, $L/2$, $3L/4$, L, $3L/2$ were prepared, where L is the period of the lamellar morphology in the bulk ($L = 762$ Å [2]). Ellipsometry and X-ray reflectivity were used to determine the specimen thicknesses. Note that while there is an electron density difference between PS and PMMA, this difference is much smaller than that between air/P(D-S-b-MMA) and P(D-S-b-MMA)/Si interfaces, which dominated the X-ray reflectivity curve. X-ray reflectivity measurements on all the five specimens yielded relatively simple profiles with oscillations characteristic of only the film thickness. The roughness at both air/polymer

Table 1. Sample thicknesses

Label	Desired thickness (Å)	Thickness by ellipsometry (Å)	Thickness by X-ray reflectivity (Å)	Thickness by neutron reflectivity (Å)
$L/4$	190	188	215	209
$L/2$	381	400	406	381
$3L/4$	571	568	602	555
L	762	734	740	726
$3L/2$	1143	1320	1365	1299

and polymer/Si interfaces was about 6 Å, assuming a Gaussian distribution of deviations about a mean value. The thicknesses determined by ellipsometry, X-ray reflectivity and neutron reflectivity are shown in Table 1. The values agree with one another to within less than 10%.

X-ray photoelectron spectroscopy (XPS) experiments on specimens prepared in identically the same way and annealed under the same conditions reveal the relative concentrations of both components as a function of depth over the first 75 Å from the air/copolymer interface, with the O_{1s} and C_{1s} signals serving as signatures of the PMMA and PS portions of the diblock copolymer. No variation

Table 2. X-ray photoelectron spectroscopy results and average compositions

Thickness	$(f_s)_{NR}$[a]	$(f_s)_{XPS}$[b]	$\dfrac{(\phi_{PS,ave})_{NR}}{\phi_{PS,nominal}}$	Percentage deviation
$L/4$	1.0	1.0	0.58/0.57	1.8
$L/2$	1.0	1.0	0.59/0.57	3.5
$3L/4$	1.0	1.0	0.62/0.57	8.8
L	0.54	0.54	0.62/0.57	8.8
$3L/2$	1.0	1.0	0.57/0.57	0.0

[a]) Surface composition of D-PS revealed by neutron reflectivity
[b]) Surface composition of D-PS integrated over the top 75 Å

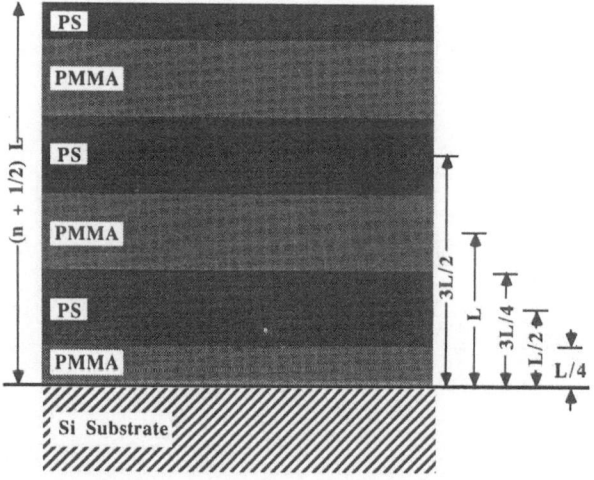

Fig. 1. Schematic diagram of the ordered lamellar microdomain morphology found in thin films of symmetric diblock copolymers of P(S-b-MMA). Specific details can be found in ref. [2]. To the right are the nominal thicknesses of the specimen investigated in this study

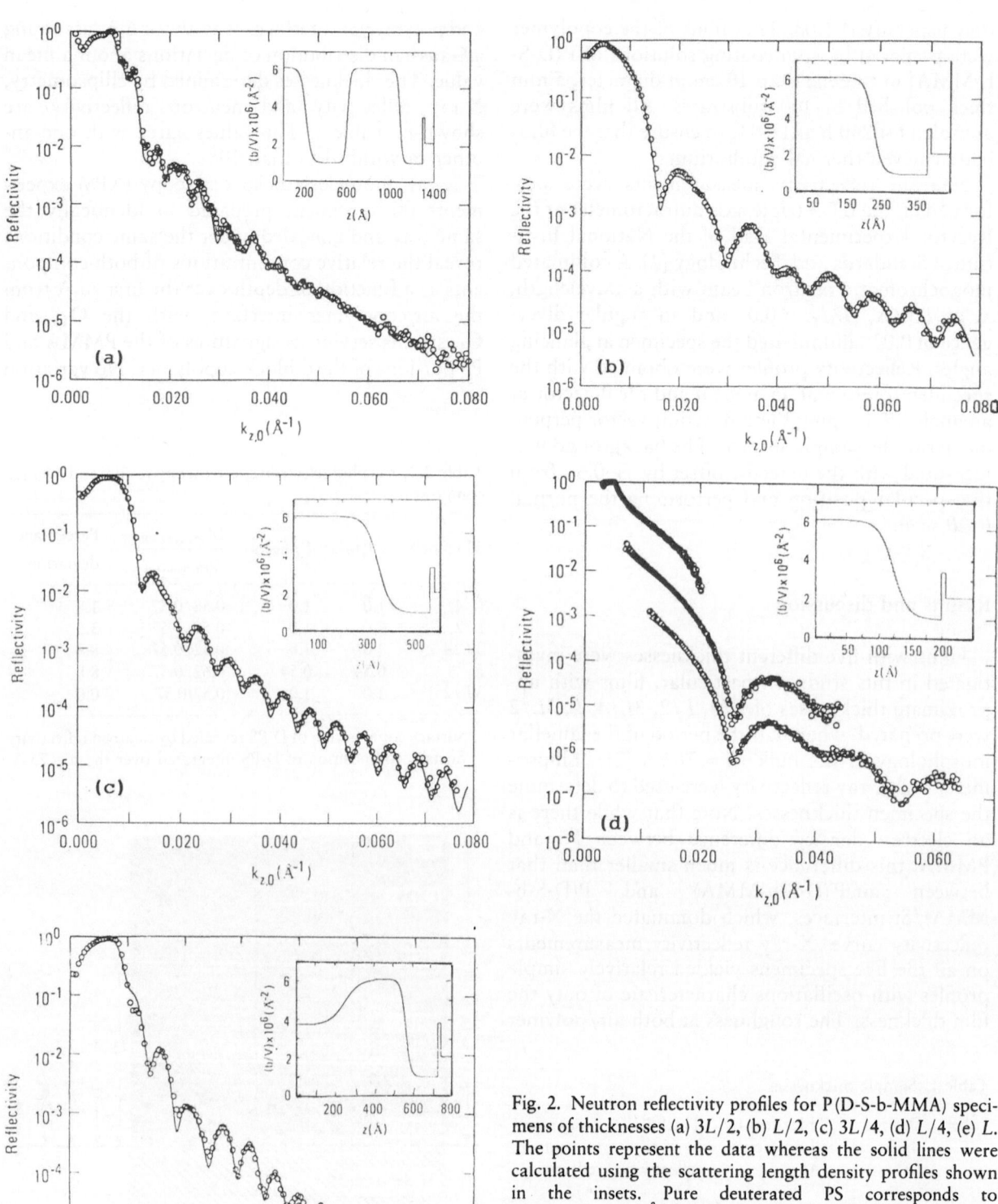

Fig. 2. Neutron reflectivity profiles for P(D-S-b-MMA) specimens of thicknesses (a) $3L/2$, (b) $L/2$, (c) $3L/4$, (d) $L/4$, (e) L. The points represent the data whereas the solid lines were calculated using the scattering length density profiles shown in the insets. Pure deuterated PS corresponds to $b/V = 6.1 \cdot 10^{-6} \text{Å}^{-2}$, whereas pure hydrogenous PMMA corresponds to $b/V = 1.0 \cdot 10^{-6} \text{Å}^{-2}$. Zero in the inset indicates the air surface. A silicon oxide layer of about 10–20 Å was necessary to be included at the Si side of the profile. In (d), the three profiles were obtained on POSY-II at IPNS, Argonne National Laboratory, at three different angles of incidence. They are offset from one another by factors of 10

in concentration with depth was observed over the top 75 Å of the specimens. The results of XPS are shown in Table 2 as the surface composition of styrene integrated over the first 75 Å. For all the sample thicknesses investigated, except the specimen with thickness L, only PS was observed in the first 75 Å from the surface. This result alone states that the copolymer films are microphase-separated. For the L-thick specimen, the volume fraction of PS at the surface was found to be 0.54, that corresponds to the composition of the diblock copolymer and indicates that at the surface the copolymer is phase-mixed. XPS has provided some very important pieces of information prior to the use of any modelling (necessary for the neutron reflectivity studies) to which the neutron reflectivity results must conform.

The morphology of thin diblock copolymer films on Si substrates as determined from previous neutron reflectivity studies is shown schematically in Fig. 1. The total thickness of the specimens is given by $(n + 1/2)L$, where n is an integer. This morphology is produced because of the special affinity of PMMA to the Si substrate and the lower surface energy of the PS. Two of the thicknesses investigated in this study, namely $3L/2$ and $L/2$, correspond to the cases where n is 1 or 0, respectively. Consequently, these films should yield reflectivity results which are relatively easy and straightforward to interpret. The neutron reflectivity profiles for all five specimens as a function of $k_{z,0}$, the component of the incident neutron momentum perpendicular to the surface ($k_{z,0} = (2\pi/\lambda)\sin q$), are shown in Fig. 2 together with the reflectivity curves calculated using the scattering length density profiles shown in the insets. In (d), the three profiles were obtained on POSY-II at IPNS, Argonne National Laboratory, at three different angles of incidence. They are offset from one another by factors of 10. Details of the time-of-flight neutron reflectivity experiments can be found elsewhere [6, 7]. A very good agreement between the measured and the calculated reflectivity profile over the entire $k_{z,0}$ range and almost six orders in reflected intensity is observed. An extended discussion and details of the profiles can be found elsewhere [8]. The main points of the profiles can be summarized as follows.

For the case of the $L/2$ specimen, a layer of pure PS is found to be located on top of a layer of pure PMMA, with an interface that can be described by a hyperbolic tangent function with an effective width, a_1, of 50 Å, in perfect agreement with thicker specimens [2], i.e., for $n = 0$ ($L/2$ specimen) an unperturbed morphology is found.

For the $3L/2$ specimen the reflectivity curve is modulated by frequencies characteristic of the thickness of the layers formed in the multilayer structure of the copolymer. The scattering length density profile shows alternating PS and PMMA domains, with interfaces in between the microdomains, similar to the ones for multilayered specimens. Therefore, the $n = 1$ ($3L/2$) case is shown to exhibit a morphology reminiscent of the one for a multilayered specimen.

The $3L/4$ specimen shows a phase-separated layered morphology with a PS layer on top of a PMMA layer. A nearly 50% increase in the microdomain sizes over that found for the $L/2$ specimen is observed together with an interfacial width of about 100 Å, much larger than that seen for the $L/2$ specimen and in the bulk. The retention of the bilayered morphology mandates that the number of copolymer chains crossing the interface has increased and that the extension of the copolymer chains has also increased. Besides, the system sacrifices unfavorable contacts at the interface in order to prevent further stretching.

The $L/4$ data show again a distinct bilayered morphology evident by the two different frequencies of oscillations in the reflectivity curve. A pure PS layer is located on top of a pure PMMA layer separated by an interface of about 50 Å, identical to the one for the $L/2$ specimen. From these observations, a possible picture of the copolymer chains is that they remain perpendicular to the interface and the extension of the chains at the interface is reduced in comparison to the $L/2$ specimen. Even in this highly confined case, the special interactions of PMMA with Si and PS with air are sufficient for the retention of the microphase-separated morphology.

Finally, the profile for the L specimen is much more complex. The X-ray photoelectron spectroscopy data on this specimen show a 0.54 volume fraction of PS at the first 75 Å of the specimen. (Note that XPS data for all the other specimens verified that the volume fraction of PS at the top 75 Å is 1.0.) It is apparent from the scattering length density profile that the copolymer initially forms a bilayer at the substrate (PMMA below PS, similar to the $L/2$ specimen). Then, the rest of the copolymer is essentially in contact with a soft PS substrate. The copolymer then faces a situation where PS preferentially segregates to both the top and the bottom surfaces. Faced with this situation, the diblock copolymer does not phase-separate, but rather remains phase-mixed.

As a test of the internal consistency of all the models presented, the average PS volume fraction is computed and compared to the actual PS content in the sample in Table 2. The average composition of the models is to within 9% of the nominal value.

In summary, the constraints placed on a copolymer film by the thickness of the specimen can produce severe perturbations of the morphology. These perturbations are strongly influenced by the interactions of the copolymer blocks with the air and the substrate interfaces. It is possible to induce variations in the chain extension and in the interfacial mixing and to force a normally microphase-separated copolymer into a phase-mixed state.

Acknowledgements

We thank D. C. Miller of the IBM Almaden Research Center for performing the X-ray photoelectron spectroscopy measurements. This work was partially supported by the Department of Energy, Office of Basic Energy Sciences, under grant number DE-FG-03-88ER45375.

References

1. Anastasiadis SH, Russell TP, Satija SK, Majkrzak CF (1989) Phys Rev Lett 62:1852–1855
2. Anastasiadis SH, Russell TP, Satija SK, Majkrzak CF (1990) J Chem Phys 92:5677–5691
3. Menelle A, Russell TP, Anastasiadis SH, Satija SK, Majkrzak CF (1992) Phys Rev Lett 68:67–70
4. Russell TP, Menelle A, Anastasiadis SH, Satija SK, Majkrzak CF (1991) Colloid Polym Sci (this issue)
5. Fredrickson GH (1987) Macromolecules 20:2535–2542
6. Russell TP (1990) Mater Sci Rep 5:171–271
7. Russell TP, Anastasiadis SH, Menelle A, Felcher G, Satija SK (1991) Macromolecules 24:1575–1582
8. Russell TP, Menelle A, Anastasiadis SH, Satija SK, Majkrzak CF (1991) Macromolecules 24:6263–6269

Received October 15, 1991;
accepted May 28, 1992

Authors' address:

Dr. Spiros H. Anastasiadis
Foundation for Research and Technology-Hellas
Institute of Electronic Structure and Laser
P.O. Box 1527
711 10 Heraklion Crete, Greece

Progress in Colloid & Polymer Science

Progr Colloid Polym Sci 91:93–96 (1993)

Short-time dynamics of polymer diffusion across an interface

G. Reiter[1]) and U. Steiner[2])

[1]) Dept. Materials Science and Engineering, University of Illinois, Urbana, USA
[2]) Department of Polymer Research, Weizmann Institute, Rehovot, Israel

Abstract: We investigated diffusion between deuterated and protonated polystyrene films at short times using neutron reflectometry (NR) and nuclear reaction analysis (NRA). We covered six orders of magnitude in time and identified two time regimes of diffusion with a crossover at the reptation time τ_{rep}. Below τ_{rep} the interfacial profile can be described by two superposed error functions attributed to the Rouse-type mobility and reptation. The contribution due to reptation shows a discontinuity at the position of the interface. There are some strong indications for an enhanced concentration of chain ends at the interface. The reliability of the results is supported by the agreement of the three values of τ_{rep} obtained by the use of the two complementary techniques.

Key words: Polymer diffusion – reptation – Rouse model – neutron reflection – nuclear reaction analysis

Introduction

In contrast to diffusion of small molecules in simple liquids motion in polymeric melts is constrained. The connectivity of monomers demands correlated movements. This behavior is described well by the Rouse model [1]. If the molecules are long enough they are entangled restricting movements to a limited region called tube. The chain-like molecules can only move along this tube (= reptation) [2]. The two constraints are relevant on different time scales which is expressed by characteristic time regimes of polymer diffusion at short times [3].

We examined these problems by measuring the interfacial profile between two thin polymer films as a function of annealing time applying two complementary techniques, neutron reflectometry (NR) [4] and $^2H(^3He, ^4He)^1H$ nuclear reaction analysis (NRA) [5]. NR can resolve interfacial profiles up to 10–20 nm, whereas NRA can detect interfacial broadening only if the width is larger than approximately 10 nm. The geometry of thin films is advantageous because it enables us to detect the tube constraint not only by measuring characteristic

time behavior of the growth of the interfacial region. Moreover, the tube constraint is indicated in the shape of the interfacial profile. All chain segments positioned at the interface (coordinate of depth $z = 0$) can move across the interface up to a distance proportional to the tube diameter D_T. To move further, however, the molecules have to reptate and, thus, initially only ends, which represent a small fraction of segments, can diffuse further than D_T. The resulting interfacial profile, therefore, is discontinuous at $z = 0$, when regarding the contribution due to reptation only [6].

We used deuterated $(M_w = 752\,000\ \text{g mol}^{-1}$, $M_w/M_n = 1.2)$ and protonated $(M_w = 660\,000\ \text{g mol}^{-1}$, $M_w/M_n = 1.1)$ polystyrene for the present experiments. The degrees of polymerization are almost equal and high enough for chains to be entangled, but low enough for molecules to be still miscible. Single films were prepared by spincoating and a double layer is prepared by floatation technique. The samples are annealed under vacuum, quenched down to room temperature and measured by NR for several times. Thus, we cover a large range in annealing times with a single sample. For NRA we had to use a new sample for each data point, because the ion beam partially destroys the sample. Different annealing temperatures are reduced to the reference temperature of 120 °C by a shift factor given by the WLF equation [7]. As all annealing times are longer than the Rouse time of our system, the only important process for diffusion is reptation. In this case the WLF equation is expected to hold. For shorter times different processes are also relevant and this might have some influence on the WLF parameters.

All resulting data concerning the interface are corrected for initial roughness. Further experimental details are published elsewhere [8].

The analysis of the NR spectra clearly resolved that the interfacial profile cannot be described by a simple error function characteristic for the Fickian diffusion of small molecules [9]. A nearly perfect fit of the data was achieved using the superposition of two weighted error functions according to the following equation for the refractive index n [4] as a function of z:

$$n(z) = \Delta n[(1 - p)\,\mathrm{erfc}\,(z/\sigma_c) + p \cdot \mathrm{erfc}\,(z/\sigma_t)],$$

$$(\sigma_c < \sigma_t)\,, \tag{1}$$

with Δn the difference in refractive index of the deuterated and the protonated layer, σ_c and σ_t the parameters of the error functions and p the weight of the second error function. erfc is the complementary error function.

The parameters can be interpreted in terms of the Rouse and the reptation model. σ_c is attributed to the Rouse-type mobility of all segments within the tube. Thus, the maximum value of σ_c is determined by D_T. From the experiments we observe σ_c to be almost constant (2–3 nm) with time in the time range measured by NR. The only exception is the jump from zero to 2 nm after the very first annealing which, however, was longer than τ_e, the characteristic time for the first influence of the tube on the diffusion process [3]. This broadening of the interface at very short times was also observed by others [10, 11].

With increasing time the "tails" of the profile get progressively larger, both in length and in relative contribution to the profile expressed by the parameters σ_t and p, respectively. According to the reptation model this can be explained by molecules moving out of their original tubes. Thus, the interface gets broader. The contribution due to reptation increases because at the beginning only ends

can leave the tube but later on also inner segments will follow. Assuming random distribution of chain ends the time dependence of the parameter p is expected to be $t^{1/2}$ [6, 12]. Our experimental results however yield $p \sim t^{0.22}$. A possible explanation for this discrepancy would be that the ends of all molecules within a surface layer of thickness proportional to the radius of gyration are attracted to the interface. Such an enhanced concentration of ends at the interface is predicted to yield $p \sim t^{1/4}$ [13] which is close to our experimental results. There are, however, other possible influences on the time dependence of p like nonequilibrium chain conformations at the interface (possibly enhanced by the spincoating process) or different mobility of parts of the chain compared to the center of mass mobility. A summary of the exponents for the time dependences of the different parameters is given in Table 1.

The interfacial width W as a function of time is plotted in Fig. 1. The inset shows σ_c, σ_t and p as a function of time. We calculated W_NR from the three parameters of our model according to the following equation:

$$W_\mathrm{NR} = [(1 - p)\sigma_c^2 + p \cdot \sigma_t^2]^{1/2}\,. \tag{2}$$

The time dependence of W_NR is $t^{0.24}$ which is very close to the expected value of 1/4 for this time regime in the *bulk* [3]. This again is a hint for an enhanced concentration of chain ends at the interface. A random end distribution would yield a different exponent because diffusion of a molecule across the interface would be hindered and, thus, delayed as long as its end has not crossed the interface.

We were also able to determine the reptation time τ_rep. Having shown that we are dealing with the Fickian diffusion in the time range covered by the NRA data ($W_\mathrm{NRA} \sim t^{0.53}$) we are able to calculate a diffusion coefficient and accordingly τ_rep [3]. The extrapolation of the parameter p to $p = 1$ gives also a value for τ_rep and the crossover of the extrapolated lines of $W_\mathrm{NR}(t)$ and $W_\mathrm{NRA}(t)$ provides a third measure for τ_rep. The results of τ_rep and the

Table 1. Power law exponents $x \sim t^\alpha$

Parameter x	σ_c	σ_t	p	W_NR	W_NRA
Exponent α	0.07	0.17	0.22	0.24	0.53

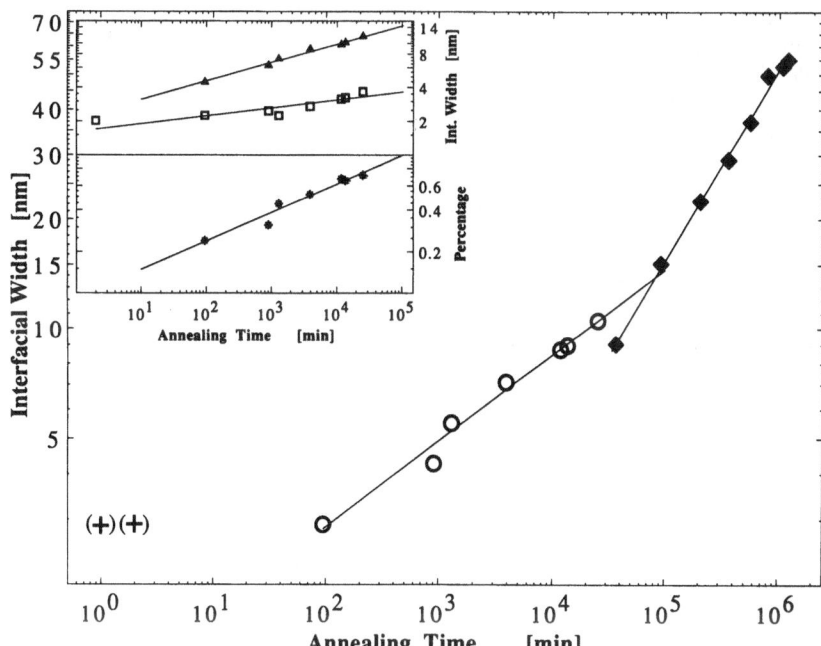

Fig. 1. Double logarithmic plot of the interfacial width $W(t)$ vs. time. NRA data (■) and NR data (○). The values are corrected for initial roughness of the interface. The crosses in brackets represent the uncorrected unannealed (arbitrarily set to $t = 1$ min) and the briefly annealed sample ($t = 2$ min), to show the lower limit of $W(t)$ which could be measured. The inset shows the time dependence of p (*), σ_c (□) and σ_t (▲). Exponents are given in Table 1

Table 2. Reptation time τ_{rep} and interfacial width $W(\tau_{rep})$

	$p \rightarrow p = 1$	Diffusion coefficient (NRA)	Crossover NR–NRA
τ_{rep}(min)	$1.04 \cdot 10^5$	$6.85 \cdot 10^4$	$8.2 \cdot 10^4$
$W(\tau_{rep})$ (nm)	14.5	12.7	13.9

corresponding values of $W(\tau_{rep})$ are listed in Table 2. The good agreement of the values achieved by three different techniques supports the model used to describe the NR data. The values of $W(\tau_{rep})$ compare well to the value one expects in the bulk ($W(\tau_{rep}) \simeq 14$ nm [3, 8]). This is a further indication that there is no delay of diffusion across the interface supporting the assumption of an enhanced concentration of chain ends at the interface.

In conclusion, we showed that polymer diffusion at short times cannot be described by the Fickian diffusion but can be explained by reptation arguments. In particular, the shape of the interfacial profile as revealed by NR can be expressed in terms of the Rouse and the reptation model. The center of the profile remains rather steep, whereas the tails increase with time. The contribution ascribed to reptation is discontinuous at $z = 0$. There is

a strong indication that in the present system the chain ends are attracted to the surface of the films. The reliability of the results is reinforced by the agreement of τ_{rep} and $W(\tau_{rep})$ determined in three independent ways.

Acknowledgement

We are grateful to J. Klein and M. Stamm for the support of this work. We have greatly benefited from discussions with P.G. dé Gennes, J. F. Joanny, A. Silberberg, J. Harden, A. Halperin, S. Hüttenbach and J. Reiter. We thank E. Eiser and B. Derichs for their help during the experiments. Financial support by ESF, BMFT, GIF, Minerva and the US-Israeli Binational Science Foundation is acknowledged.

References

1. Rouse PE (1953) J Chem Phys 21:1272
2. De Gennes PG (1971) J Chem Phys 55:572; De Gennes PG (1979) Scaling Concepts in Polymer Physics. Cornell University Press, Ithaka, NY
3. Doi M, Edwards SF (1986) Theory of Polymer Dynamics. Oxford University Press, Oxford
4. Reiter G, Hüttenbach S, Foster M, Stamm M, Fresenius J (1991) Anal Chem 341:284–288; Russell TP (1990) Materials Sci Rep 5:173
5. Chaturvedi UK, Steiner U, Zak O, Krausch G, Schatz G, Klein J (1990) Appl Phys Lett 56:1228

6. De Gennes PG (1980) C R Acad Sci Paris serie B:219–222; De Gennes PG (1989) C R Acad Sci Paris 308 serie II:13–17
7. Tassin JF, Monnerie L (1988) Macromolecules 21:1846
8. Reiter G, Steiner U (1991) J Phys II (France) 1:659–671 Steiner U, Reiter G, Eiser E, Klein J, to be published
9. Crank J (1975) The Mathematics of Diffusion 2nd ed. Clarendon Press, Oxford
10. Karim A, Mansour A, Felcher GP, Russell TP (1990) Phys Rev B 42:6846
11. Stamm M, Hüttenbach S, Reiter G, Springer T (1991) Europhys Lett 14:451–456
12. Prager S, Tirell M (1981) J Chem Phys 75:5194; Wool RP, Yuan BL, McGarel OJ (1989) Polym Eng Sci 29:1340
13. De Gennes PG (1988) C R Acad Sci Paris 307 serie II:1841–1844; De Gennes PG, private communication

Received January 14, 1992;
accepted May 5, 1992

Authors' address:

Günter Reiter
Dept. Materials Science and Engineering
University of Illinois
Urbana, Il 61801, USA

Progress in Colloid & Polymer Science

Progr Colloid Polym Sci 91:97–100 (1993)

The ordering of thin films of symmetric diblock copolymers

T. P. Russell, A. Menelle, S. H. Anastasiadis, S. K. Satija[1], and C. F. Majkrzak[1]

IBM Research Division, Almaden Research Center, San Jose, California, USA
[1]) National Institute of Standards and Technology, Reactor Radiation Division, Gaithersburg, USA

Abstract: The ordering of thin films of symmetric diblock copolymers of polystyrene and polymethylmethacrylate has been investigated by neutron reflectivity. It is shown that the order–disorder transition temperature depends strongly on the film thickness, increasing as the film thickness decreases. The transition is shown to lose all its first-order characteristics in the thin films. For films thinner than ~ 2000 Å it is not possible to obtain the copolymer films in the disordered state at temperatures below the decomposition temperature. A transition is observed where the copolymer goes from a partially to a fully ordered state. For thick specimens, this transition corresponds to the bulk order–disorder transition temperature. However, for thinner specimens this transition temperature depends on the film thickness in a power law manner.

Key words: Block copolymer – surface ordering – thin films – neutron reflectivity

Introduction

Symmetric diblock copolymers undergo a transition from a disordered state to an ordered, lamellar microdomain morphology at a temperature T_{ODT}, where $\chi N = 10.5$ [1]. Here, χ is the Flory–Huggins segmental interaction parameter, which varies inversely with temperature, and N is the total number of segments in the copolymer chain. Neutron or X-ray scattering from a diblock copolymer in the disordered state exhibits a maximum which occurs at a scattering vector that scales inversely with the radius of gyration of the copolymer and which has a magnitude which depends on the temperature difference from T_{ODT}. As T_{ODT} is approached, the intensity of the maximum diverges. Fredrickson and Helfand [2] have shown that concentratioin fluctuations are of critical importance and that the order–disorder transition in the bulk is first-order.

In the vicinity of an interface, however, a preferential interaction of the blocks with the interface can induce an ordering in a disordered diblock copolymer. Mean field arguments [3] predict that, as a function of distance z from the interface, the composition of either component will vary in an exponentially damped, sinusoidal manner. Far from T_{ODT}, this behavior has been observed experimentally and has been shown to be in very good agreement with the simple mean field arguments [4]. Theoretically, as T_{ODT} is approached, the decay length increases in a manner that is similar to that seen in bulk films. Herein, a neutron reflectivity study of the ordering of thin films of symmetric diblock copolymers is discussed as a function of temperature and film thickness. It has been found that the order–disorder transition in thin diblock copolymer films is no longer first order in nature and that a second transition from a partially ordered to a fully order state occurs. The temperature of both the transitions is found to depend strongly on the film thickness.

Experimental

The copolymer used in this study was a symmetric diblock copolymer of polystyrene and polymethylmethacrylate where the latter block was perdeuterated. This was denoted as P(S-b-d-MMA) and had an M_w of 29700, with

$M_w/M_n = 1.1$. The fractioin of PS in the copolymer was 0.5. Thin films of the copolymer were prepared by spin-coating solutions of P(S-b-d-MMA) in toluene onto 10 cm diameter, 5 mm thick Si. The thickness of the films ranged from ~ 900 Å to 1.2 μm. The concentration of P(S-b-d-MMA) was used to vary the film thickness. The films were initially dried at 80 °C under vacuum for 48 h to remove the residual solvent. The samples were then annealed at the desired temperature for at least 12 h, quenched to room temperature and then measured. The annealing temperature was varied in a nonsystematic manner over the desired temperature range to eliminate systematic errors.

Neutron reflectivity measurements were performed on a neutron reflectometer at Beamline BT7 in the Reactor Hall of the National Institute of Standards and Technology. A monochromatic beam of neutrons with a wavelength, λ, of 2.35 Å was employed. Slits collimated the neutron into a line with dimensions of 75 μm (horizontal) by 2.5 cm (vertical) with respective divergences of 0.01 ° and 2 °. The horizontal width of the incident beam was increased with increasing angle to enhance the flux on the specimen. In these studies $\Delta\theta/\theta$ varied as a function of angle and was taken into account in the calculation of the reflectivity profiles. Measurement of the reflectivity 0.3 ° off the specular position provided a measure of background which was subtracted from the specularly reflected intensity.

Reflectivity profiles were calculated by assuming a functional form of the variation in the scattering length density in the specimen. This was then reduced to a histogrammatic representation of (b/V) as a function of depth z where the step size of the histogram did not necessarily remain constant. Details of the manner in which the reflectivity is calculated have been presented elsewhere [5, 6]. The concentration of PS as a function of depth z, in a specimen of thickness E, was assumed to be given as

$$\phi_{PS}(z) = \max\left\{\phi_A e^{-z/\xi} \cos\left(\frac{2\pi z}{\bar{L}}\right),\right.$$
$$\left.\phi_S e^{-(E-z)/\xi}\cos\left[\frac{2\pi(E-z)}{\bar{L}}\right]\right\} + \bar{\phi}_{PS},$$
$$(1)$$

where ϕ_A and ϕ_S are the excess volume fractions of PS at the air and substrate interfaces, respectively,

$\bar{L}$ is the average period, ξ is the decay length and $\bar{\phi}_{PS}$ is the average concentration of PS in the copolymer. Measurements of ϕ_A by X-ray photoelectron spectroscopy, $\bar{L}$ by small-angle neutron scattering, and $\bar{\phi}$ from the copolymer composition and assuming $\phi_A = -0.5$ (due to the strong interactions of PMMA with the substrate) reduced the calculation of the reflectivity to two parameters, namely, ξ and the surface roughness. These were varied to produce the best fit to the experimental data.

Results and discussion

A typical set of neutron reflectivity profiles as a function of the neutron momentum normal to the film surface, $k_{z,0} = (4\pi/\lambda)\sin\theta$, where θ is the grazing angle of incidence, is shown in Fig. 1 at three different temperatures. The reflectivity profiles are characterized by a strong first-order reflection at ~ 0.02 Å^{-1}, followed by higher-order reflections at ~ 0.036 and ~ 0.056 Å^{-1}. As the temperature increases, the first-order reflection broadens and reduces in intensity and the higher-order reflections become weaker and ill-defined. For much thicker specimens [7] the higher-order reflections are completely lost and only a slight ridge in the reflectivity is seen. The (b/V) profiles shown in the inset, derived from Eq. (1), were used to calculate the reflectivity profiles shown as the solid lines in the figure. As can be seen, the agreement between the calculated and the measured reflectivity profiles is quite good over the entire $k_{z,0}$ range. While the (b/V) profiles represent only one possible model, the essential physics is captured in these profiles.

An examination of these (b/V) profiles shows that for this specimen a truly disordered state is never achieved, even at 210 °C. Considering that the order–disorder transition temperature, T_{ODT}, occurs at ~ 157 °C and $\bar{L} \sim 168$ Å, these results show the strong influence that the interactions of the P(S-b-d-MMA) with the air and substrate interfaces have on the ordering and the extent to which the ordering propagates into the bulk of the specimen. Increasing the specimen thickness results in (b/V) profiles which decay to $\bar{\phi}_{PS} = 0.5$ at lower temperatures; hence, it is possible to attain disordered films when the decay lengths are much less than E.

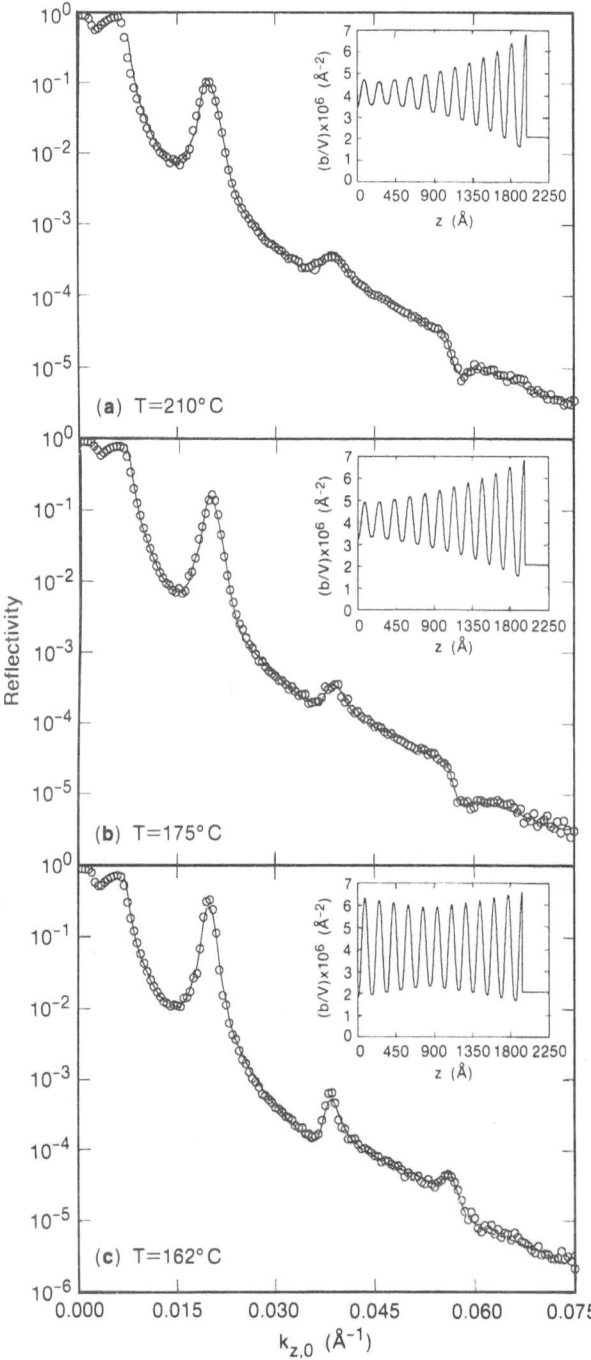

Fig. 1. Neutron reflectivity profiles for a P(S-b-d-MMA) film 1962 Å in thickness as a function of the neutron momentum $k_{z,0}$ at the temperatures indicated. The scattering length density profiles shown in the inset were used to calculate the reflectivity profiles shown as the solid lines

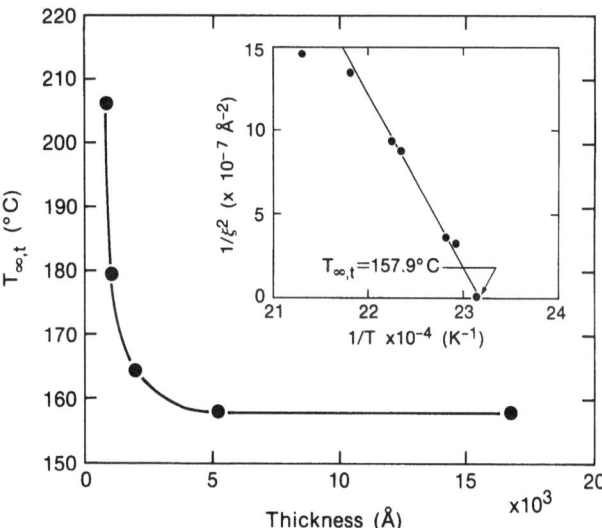

Fig. 2. The temperature at which the P(S-b-d-MMA) films underwent a transition from a partially to a fully ordered state as a function of the film thickness. In the inset is shown the inverse square of the decay length as a function of the inverse temperature from which the transition temperature was extrapolated

From the (b/V) profile ξ can be obtained. Shown in the inset in Fig. 2 is a plot of ξ^{-2} as a function of T^{-1}. Curvature is seen in these data over the entire temperature range, but the data at lower temperatures can be extrapolated to a point where the copolymer film is fully ordered, i.e., at $T_{\infty,t}$ $\xi \to \infty$. This represents a transition from a partially to a fully ordered state, a transition not previously seen in diblock copolymers. From these data and those of films with different thicknesses, $T_{\infty,t}$ was plotted as a function of the film thickness. As snown in Fig. 2, for thick specimens $T_{\infty,t}$ extrapolates to the value of T_{ODT} seen for bulk specimens. However, as the film thickness decreases, $T_{\infty,t}$ increases dramatically. Thus, these data show that, for thin films (less than $\sim 2 \cdot 10^3$ Å), it is most difficult to get a diblock copolymer film that is anything but fully ordered.

It is important to note that the order–disorder transition in diblock copolymer films is seen only when the decay length is finite and has a magnitude much less than the total film thickness. Through this transition ξ varies smoothly, with no dramatic changes. This indicates that the order–disorder transition, which is first-order in the bulk, has lost its first-order nature in the thin films.

Acknowledgements

This work was supported by the Department of Energy, Office of Basic Energy Sciences Grant No. DE-FG-03-88ER45375.

References

1. Leibler L (1980) Macromolecules 13:1602–1617
2. Frederickson GH, Helfand EJ (1987) J Chem Phys 87:697–705
3. Frederickson GH (1987) Macromolecules 20:2535–2540
4. Anastasiadis SH, Russell TP, Satija SK, Majkrzak CF (1989) Phys Rev Lett 62:1852–1855
5. Als-Nielsen J (1987) In: Schommers N, von Blanckenchagen P (eds) Structure and Dynamics of Surfaces II. Springer, Berlin, pp. 181–222
6. Russell TP (1990) Mat Sci Rep 5:171–282
7. Menelle A, Russell TP, Anastasiadis SH, Satija SK, Majkrzak CF (1992) Phys Rev Lett 68:67–70

Received January 25, 1992;
accepted June 2, 1992

Authors' address:

T. P. Russell
IBM Research Division
Almaden Research Center
650 Harry Road
San Jose, CA 95120-6099, USA

Progress in Colloid & Polymer Science   Progr Colloid Polym Sci 91:101–104 (1993)

Organized structures in diblock copolymer films of polystyrene and poly-para-methylstyrene

M. Stamm[1]), A. Götzelmann[1]), K. H. Gießler[2]), and F. Rauch[2])

[1]) Max-Planck-Institut für Polymerforschung, Mainz, FRG
[2]) Institut für Kernphysik, Universität Frankfurt/Main, FRG

Abstract: A diblock copolymer of poly(styrene-block-para-methylstyrene) (PS-b-PMS) is investigated near its microphase separation temperature with respect to the formation of organized structures in thin films. A contrast between components is achieved by deuteration of the PS-block, and the surface-induced order is detected by neutron reflectometry and ^{15}N nuclear reaction analysis. While the PMS component is always enriched at the surface, an exponentially decaying lamellar order of alternating layers of PS and PMS parallel to the surface is observed depending on annealing time. Those observations are discussed with respect to the influence of surface and substrate on the development of organized structures in a block copolymer melt.

Key words: Diblock copolymers – thin films – order – neutron reflectometry – nuclear reaction analysis

Introduction

The bulk properties of polymer blends have been investigated by various techniques and also the spinodal behaviour of polystyrene (PS) and poly-para-methylstyrene (PMS) is well known [1]. The system shows an upper critical solution temperature (UCST). Depending on the molecular weight and composition it is immiscible at low and miscible at higher temperatures. The diblock copolymer PS-b-PMS has been reported to show a similar behaviour [2] and, in particular, the predictions of Leibler [3] are nicely obeyed showing that $(N\chi)_s \sim 2$ for homopolymer blends and $(N\chi)_s \sim 11.5$ for the investigated diblock copolymers. N is the degree of polymerisation, χ the Flory–Huggins interaction parameter, and the index s denotes the values at the spinodal. This blend can, thus, be taken as a model system for spinodal decomposition.

The behaviour of blends at the surface is far less investigated. Because of the difference in surface energy the component with the lower surface en-

ergy shows generally a tendency for surface enrichment. In specific cases of homopolymer blends near the critical decomposition temperature this may lead to a surface-directed spinodal decomposition process with a concentration modulation perpendicular to the surface, which changes with annealing time and extends over a micrometer scale [4]. Surface-directed order is also known for block copolymers, where e.g. the lamellar morphology in the two-phase region is ordered parallel to the surface. This has been observed by electron microscopy [5], dynamic secondary-ion mass spectroscopy [6] and neutron reflectometry [7]. Block copolymers investigated are usually in the strong segregation limit. In a recent neutron reflectometry experiment the surface-induced order in the one-phase region near the critical temperature T_c has been studied [7] showing a surface-induced order with decaying amplitude. Such a behaviour is also described by theory [8] which predicts an exponentially damped oscillation normal to the surface induced by the preference of one component to the surface.

To investigate the order perpendicular to the surface one needs a resolution better than the expected periodicity which is typically several nanometers. There are not many techniques available [9] which offer this resolution. Using static secondary-ion mass spectroscopy [10] we were able to determine the surface composition of the films showing a strong enrichment of PMS at the surface. With neutron reflectometry (NR) [11] and the ^{15}N nuclear reaction analysis (NRA) technique [12] we are able to also determine the concentration modulations within the film with good resolution.

Experimental

The diblock copolymer PS-b-PMS was prepared by PSS, Mainz. The PS-block is deuterated, and the total molecular weight determined by GPC on the basis of a PS-standard amounts to $M_w = 230\,000 \text{ g mol}^{-1}$, $M_w/M_n = 1.08$. The composition is $f = N_{PS}/N = 0.47$, where N_{PS} and N are the degrees of polymerization of the PS-block and the total chain, respectively. The microphase separation temperature has been determined by small-angle neutron scattering to $T_{MST} = 159\,°C$ [2, 13]. Thin films are prepared by spin coating from toluene solution. Float glass substrates ($10 \times 10 \text{ cm}^2$) are used for NR and silicon wafers ($1.8 \times 5 \text{ cm}^2$) for NRA. Samples are annealed under vacuum for various times, and the annealing temperature is measured directly on the film.

NR experiments are performed on the neutron reflectometer TOREMA [14] at KFA Jülich utilizing a fixed wavelength of 0.43 nm. NRA experiments are performed at the 7 MeV van der Graaff accelerator of the Institut für Kernphysik, Frankfurt [12] using a $4'' \times 4''$ BGO detector with a 50 mm deep-end well and an angle of incidence of $10°$ between incident ^{15}N ions and sample surface. The depth resolution of the NR technique is typically of the order of some Ångstroms, while the depth resolution of the NRA technique depends on depth. It is at the surface approximately 4 nm and at a depth of 100 nm approximately 13 nm. Further details of the experimental set-up have been described previously [12, 14].

Results and discussion

Neutron reflectivity data from a film as prepared from toluene solution and of the same film annealed for 13 h at 140 °C are shown in Fig. 1. One can directly see from the reflectivity curves that a uniform polymer layer resulting in a single periodicity of the Kiessig fringes will not be consistent with the data and that there is quite a significant change during annealing. Model fits to the data based on the matrix technique [15] reveal the scattering density profiles shown in the insert. The quantity $(1 - \varepsilon)$ is related to the neutron-scattering-length density ρ_n via the index of refraction $n = 1 - (\lambda^2/2\pi)\rho_n = \sqrt{\varepsilon}$. It is large for the deuterated PS(D)-block and small for the protonated PMS(H)-block. One, thus, clearly observes a concentration modulation of PS(D)- and PMS(H)-components of the diblock copolymer normal to the surface, which changes in wavelength, decay length and amplitude during annealing. While it is known from x-ray reflectivity experiments that the surface roughness is approximately 1 nm, the broad surface transition zone observed in the fits is indicative of a PMS(H)-layer

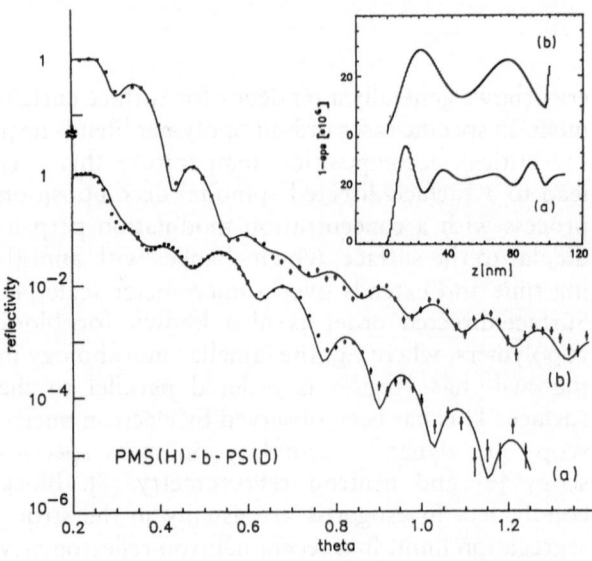

Fig. 1. Neutron reflectivity curves of an unannealed (a) and subsequently for 13 h at 140 °C annealed (b) film of a diblock copolymer PS(D)-b-PMS(H). The data are taken at a wavelength of 0.43 nm. Also shown as an insert are the corresponding scattering density profiles given through $(1 - \varepsilon)$ (see text) as obtained from fits to the reflectivity data. The fits are indicated as solid lines through the experimental points. The co-ordinate z runs normal to the surface. The surface is located at $z = 0$

at the surface, which, however, cannot be rigorously proved by NR due to the small-scattering density of PMS(H).

Static secondary-ion mass spectroscopy experiments on this blockcopolymer [10] and NRA-data shown in Fig. 2 clearly indicate the presence of a PMS(H) layer at the surface. By the resonant ^{15}N nuclear reaction with hydrogen of the sample a direct depth profile of the hydrogen distribution is obtained. This technique is, thus, complementary to NR which is mostly sensitive on the deuterium distribution. From the annealing experiments shown in Fig. 2 again the surface-induced lamellar ordering of PMS- and PS-blocks is evident. The driving force for the formation of organized structures at the surface can be seen in the lower surface energy of PMS with respect to PS causing the enrichment of PMS at the surface. The concentration of PMS at the surface is virtually 100%. Due to

the chemical connectivity the next layer has to be enriched with PS-blocks, etc.

The behaviour is quite similar to the theoretical prediction of Fredrickson [8] for the one-phase region near T_c. In particular, we do observe a thin first layer, a sinusoidal oscillation and an exponentially decaying amplitude. For a comparison of experiment and theory one has to take, however, the depth-dependent resolution function into account. During annealing the periodicity nearly doubles going from a small wavelength for the solution prepared sample to a larger wavelength for the annealed material. The largely different periodicities after preparation from solution and after annealing might be due to a significant dependence of chain organization in the bulk on sample history. This behaviour is presently not quite understood and further experiments will be necessary.

With other block-copolymers like PS-b-PMMA an apparently homogeneous film is formed after spin coating from solution which is ordered with respect to the surface during annealing [6, 7]. We clearly do observe, in conclusion, the surface-induced ordering of the PS-b-PMS diblock copolymer near the microphase separation temperature and are able to resolve the changes on a molecular level during annealing.

Acknowledgements

We acknowledge the technical help of Mrs. B. Derichs during the neutron experiments at Jülich and valuable discussions with Dr. G. Reiter, D. Endisch, Prof. E. W. Fischer and Prof. T. Springer. This work was partially supported by BMFT and DFG. The neutron experiments were performed at the KFA Jülich on the basis of a collaboration agreement.

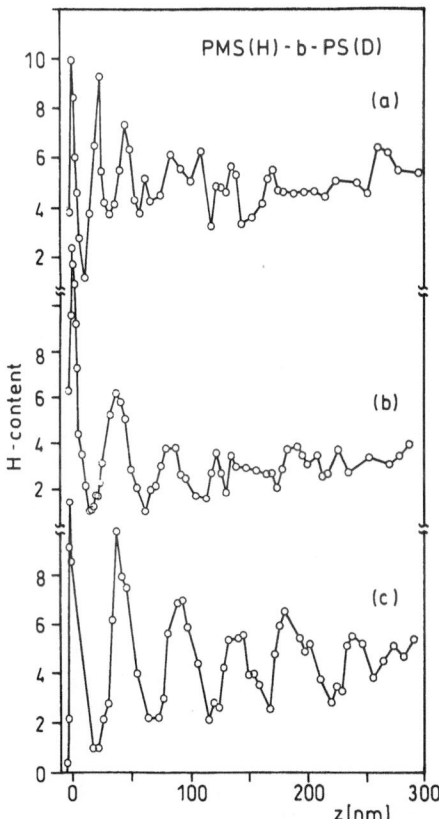

Fig. 2. Hydrogen depth profiles of PS(D)-b-PMS(H) as obtained from ^{15}N nuclear reaction analysis at different stages of annealing: (a) unannealed; (b) 20 h and (c) 60 h at 150°C annealed thin films. The solid lines are a guide to the eye. The H content is given by relative units

References

1. Antonietti M, Lang S, Sillescu H (1986) Makromol Chem Rap Commun 7:415
2. Jung WG, Fischer EW (1988) Makromol Chem Macromol Symp 16:281
3. Leibler L (1980) Macromolecules 13:1602
4. Jones RAL, Norton LJ, Kramer EJ, Bates FS, Wiltzius P (1991) Phys Rev Lett 66:1326
5. Hasegawa H, Hashimoto T (1985) Macromolecules 18:589
6. Coulon G, Russell TS, Deline UR, Green PF (1989) Macromolecules 22:2581
7. Anastasiadis SH, Russell TP, Satija SK, Majkrzak CF (1989) Phys Rev Lett 62:1852
8. Fredrickson GH (1987) Macromolecules 20:2535
9. Stamm M (1992) Adv Pol Sci 100:357
10. Affrossman S, Hindrychx F, Pethrick RA, Stamm M (1992)

In: Pireaux JJ (ed.) Polymer Solid Interfaces. Adam Hilger, Bristol, p. 337
11. Stamm M, Reiter G, Kunz K (1991) Physica B 173:35
12. Endisch D, Rauch F, Götzelmann A, Reiter G, Stamm, M (1992) Nucl Instr Meth B 62:513
13. Jung WG, private communication
14. Stamm M, Hüttenbach S, Reiter G (1991) Physica B 173:11
15. Lekner J (1987) Theory of Reflection. Martinus Nijhoff, Dordrecht

Received December 15, 1991;
accepted April 30, 1992

Authors' address:

Dr. M. Stamm
Max-Planck-Institut für Polymerforschung
Postfach 3148
D-W-6500 Mainz, FRG

Progress in Colloid & Polymer Science

Progr Colloid Polym Sci 91:105–108 (1993)

Is the distribution of entanglements homogeneous in polymer melts?

J. Bastide[1]), F. Boué[2]), E. Mendes[1]), F. Zielinski[2]), M. Buzier[1]), C. Lartigue[3]), R. Oeser[3]), and P. Lindner[3])

[1]) ICS (CNRS) Strasbourg, France;
[2]) LLB, CEN Saclay, France;
[3]) ILL, Grenoble, France

Abstract: Neutron-scattering experiments have been performed on elongated polystyrene melts consisting of relatively short labelled chains ($M < 100\,000$) dispersed in a matrix of very long ones ($M > 10^6$), which were not labelled. Important anomalies with respect to the theoretical predictions have been observed. In particular, the iso-intensity lines exhibit, in a large range of relaxation times, striking *butterfly* shapes and an unexpected orientation: the long axis of these patterns is oriented parallel to the stretching direction. As a working assumption, we propose to relate this phenomenon to a non-homogeneous arrangement of chain entanglements in the melts. As a matter of facts, it has been proposed that elongated gels should exhibit butterfly patterns (the whole solvent being labelled) as a result of a sort of anisotropic unscreening of cross-linking heterogeneities. Experiments were performed on samples synthesized specially in order to match the assumptions of the model and butterfly patterns have effectively been observed. The same type of mechanism can be driven by an heterogeneity of chain entanglement in the case of assymetrical melts: the short labelled chains may behave as a kind of polymeric solvent and accumulate, during the relaxation, in the less entangled regions (which should be also regions of lower constraint).

Key words: Polymer – melts – gels – percolation – clusters – neutron scattering

Introduction

Using the neutron-scattering technique, one can probe the correlations in polymer melts and, thus, gain some information about, for example, the mechanism of stress relaxation after applying a given elongation. The "classical" experiments [1, 2] are performed on mixtures of deuterated and non-deuterated chains of same (large) molecular weight; under these conditions (because of incompressibility) the neutron-scattering intensity is proportional to the form factor of the deuterated molecules. After imposing a given stretching ratio to the sample (at a temperature T larger than T_g), one can observe the evolution in time of the form factor of the tagged chains. In practice, one quenches, in general, the sample (at $T < T_g$) during the time of the neutron experiment and one performs series of measurements at different relaxation times t_R, i.e. after different durations of relaxation between the application of the strain and the observation. It is then possible to compare the obtained data with some theoretical curves, which are calculated after the deformation assumptions employed in the reptation model [3, 4]. The main features of the two types of curves are, in general, comparable, but, as far as we know, the agreement is never quantitative. Therefore, on the point of view of the molecular deformation at least, the description of the behaviour of polymer melts is not very precise.

The "butterfly" anomaly in asymmetrical melts

Experiments are also performed sometimes on asymmetrical systems, i.e. for example, on samples containing labelled chains significantly shorter

       Progress in Colloid and Polymer Science, Vol. 91 (1993)

than the matrix chains. A typical experiment leads to the following results [5–7]:

i) In the early stages of the relaxation process, i.e. for t_R smaller than the Rouse time of the labelled chains, a classical anisotropy is observed. The iso-intensity lines are ellipses having their long axis oriented perpendicular to the stretching direction. The scattered intensities extrapolated at $q = 0$, for **q**, respectively, parallel and perpendicular to the stretching direction, are approximately identical and equal to $I(q \rightarrow 0)$ for the isotropic sample (i.e. before deformation). Therefore, values of apparent radii of gyration in the directions parallel and perpendicular to the stretching axis (R_g^{par} and R_g^{perp}) can be extracted from the data. As expected, they are, respectively, larger and smaller than the isotropic value.

ii) As t_R is increased but kept smaller than the Rouse time of the labelled chains, the anisotropy of the elliptical iso-intensity lines and the difference between R_g^{par} and R_g^{perp} decreases. For t_R comparable with the Rouse time of the labelled chains, the iso-intensity become approximately circular again; besides, the values of both R_g^{par} and R_g^{perp} recover approximately the isotropic value R_g^{iso} (i.e. the one measured before the deformation was imposed).

iii) For t_R still larger, but smaller than the terminal relaxation time of the matrix chains (the sample still exhibits a rubbery behaviour: if it is unclamped it contracts, perhaps incompletely), a strong increase of the scattered intensity in the direction parallel to the elongation axis is observed. The iso-intensity lines adopt, at small angles, the striking shape of sort of eight, with the long axis oriented parallel to the stretching direction. These figures having an unexpected orientation have been called *butterfly patterns*. See an example in Fig. 1. It has been shown elsewhere that such an effect arises from inter-chain correlations rather than from correlations inside the same labelled molecule [5].

Butterfly patterns in stretched gels

The explanation of the butterflies in the case of elongated melts is still an open question. However,

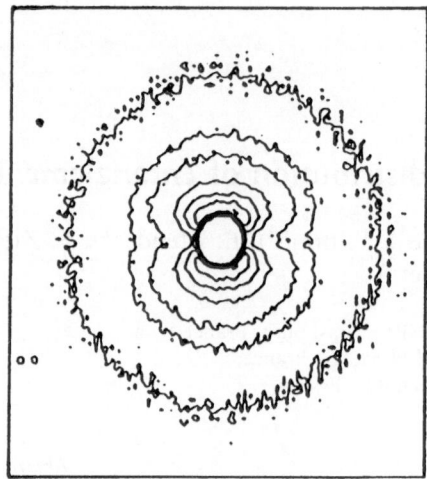

Fig. 1. Iso-intensity patterns obtained for a melt of long polystyrene chains ($M \approx 1.6 \cdot 10^6$) containing 10% of shorter labelled chains (deuterated polystyrene, $M \approx 10^5$), stretched by a factor $\lambda = L/L_0 = 3$, after a duration of relaxation of 10 min at 138 °C (step-strain experiment). $9 \cdot 10^{-3} \text{Å}^{-1} < q < 6 \cdot 10^{-2} \text{Å}^{-1}$

there are different systems, namely randomly cross-linked gels, which are expected to give rise to butterfly patterns as well, when uniaxially stretched. A model [8] has been developed on the basis of the following line of arguments.

Because of the randomness of the arrangement of junctions in space, regions more cross-linked than the gel on an average, having approximately the shape of percolation clusters, are expected to be formed. This should be valid even for a gel without significant sol fraction. In other words, the percolation model is used for the description of the disorder remaining well above the gel point. In the general case, the polymer concentration in the clusters is presumably larger than in the interstitial regions. Nevertheless, although some of the clusters can be very large, the additional scattering arising from their presence is not necessarily very important (when a contrast is established between the polymer and the solvent). This comes from the fact that intra-cluster correlations can be almost entirely screened out by intra-clusters correlations (the clusters are interpenetrated – remember that, in the true gelation problem and for the same reason of screening, the growth of the clusters cannot be studied in the reaction bath with the help of a scattering method [9]). But when the gel is elongated the situation should change. Because of the local fluctuations of the cross-linking density,

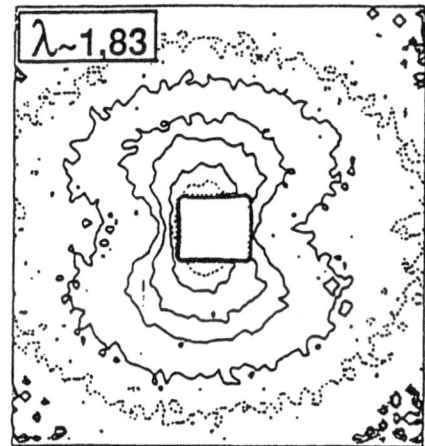

Fig. 2. Iso-intensity patterns obtained for a polystyrene gel swollen in deuterated toluene (polymer concentration $\approx 4.5\%$; close to maximum swelling) stretched by a factor $\lambda = 1.83$. The gel was prepared by statistical cross linking of a semidilute solution [11]. $6 \cdot 10^{-3} \, \text{Å}^{-1} < q < 4 \cdot 10^{-2} \, \text{Å}^{-1}$

one expects a very inhomogeneous deformation process: the more cross-linked clusters should deform less than the interstitial regions. As a result they should partly disinterpenetrate in the direction parallel to the stretching; the intra-cluster correlations are, thus, expected to be progressively unscreened; consequently, the scattered intensity at low q is expected to increase strongly. Such an effect should not exist or should remain far less important in the perpendicular direction. A simple mathematical expression of these ideas has led to the calculated iso-intensity lines having the butterfly aspect [8].

Experiments have subsequently been performed on samples synthesized specially in order to match the assumptions of the model and butterfly patterns have effectively been observed (see Fig. 2) [10–12].

Are there hidden clusters in polymer melts?

In summary, we expect the appearance of butterfly patterns in the case of randomly cross-linked gels because of an anisotropic unscreening of more cross-linked clusters, and we do observe such patterns. Then, one can inverse the reasoning and raise the question: are there also some sorts of more rigid percolation-like clusters in polymer melts, since in such systems we do observe the butterfly anomaly as well? Let us postulate this as a working assumption.

Before deformation, even if the small labelled chains are more rare in the more entangled regions (i.e. if a "contrast" is created by the partition of these labelled chains), the clusters should not be detectable on length scales larger than a certain correlation length $\xi_?$, even if their average size is larger than $\xi_?$. As in the gel problem, the reason for this "invisibility" of the clusters would be the screening of the intra-cluster correlations. After the deformation has been imposed, at small relaxation times, all the system (harder clusters and softer regions) is presumably deformed affinely up to short length scales; therefore, the clusters should still remain invisible. But at larger relaxation times (regime iii of Section II), one can expect a certain decrease of the deformation ratio of the harder clusters (and, on the other hand, a raise of the elongation of the softer parts). Then, the clusters would partly disinterpenetrate in the direction parallel to the stretching and the small chains might, after redistributing in the softer parts, play a role comparable to the solvent in the gel problem. Thus, butterfly patterns should show up.

What could be the origin of the presence of clusters

What the entanglements between chains really consist of is still debated. However, it may be conceived that their arrangement is not necessarily regular. Note, in particular, that if it were regular along the chains, then, very likely, it would not be regular in space. On the other hand, if the arrangement were regular in space it would hardly be regular along the chains. Therefore, one may imagine that some intrinsic disorder is present in the entanglement process. Then, in analogy with the situation of the randomly cross-linked networks, one may anticipate the existence of "more entangled clusters", eventually with a characteristic size larger than the average distance between entanglements.

Concluding remarks

In this short article, we wanted to present one of the possible ways for interpreting the butterfly phenomenon observed in polymer melts. Other explanations have been proposed (see refs. quoted in [7, 13]) and, at the moment, no definite conclusion has been reached about which one is the more satisfactory. Much more complete presentations of data, models and interpretations can be found in refs. [14, 15].

Acknowledgements

We are indebted to L. Leibler, H. Benoît, J. F. Joanny, C. Picot, D. Froelich and S. Panyukov for fruitful discussions.

References

1. Boué F (1982) J Phys (Paris) 43:137
2. Boué F (1987) Adv Polym Sci 82:47
3. Doi M, Edwards SF (1986) The Theory of Polymer Dynamics. Clarendon Press, Oxford
4. de Gennes PG (1979) Scaling Concepts in Polymer Physics. Cornell University Press, Ithaca, NY
5. Bastide J, Buzier M, Boué F (1988) In: Richter D, Springer T (eds) Polymer Motion in Dense Systems. Springer, Berlin, pp. 112–120
6. Boué F, Bastide J, Buzier M (1989) In: Baumgärtner A, Picot C (eds) Molecular Basis of Polymer Networks. Springer, Berlin, pp. 65–81
7. Boué F, Bastide J, Buzier, Lapp A, Herz J, Vilgis TA (1991) Coll Polym Sci 269:195–216
8. Bastide J, Leibler L, Prost J (1991) Macromolecules 23:1821
9. Daoud M, Leibler L, (1988) Macromolecules 21:1497
10. Mendes E, Lindner P, Buzier M, Boué F, Bastide J (1991) Phys Rev Lett 66:1595–1598
11. Bastide J, Mendes E, Boué F, Buzier M, Lindner P (1990) Makrom Chem Makrom Symp 40:81–99
12. Mendes E (1991) Thesis, Université Louis Pasteur, Strasbourg
13. Higgs P, to be published
14. Boué F et al. (1992) Proceedings of EPS Meeting "Physics of Polymer Networks" Alexisbad Germany, Prog Coll Pol Sci, to be published
15. Zielinski F, (1991) Thesis, Université Pierre et Marie Curie Paris VI

Received January 23, 1992;
accepted June 22, 1992

Authors' address:

J. Bastide,
ICS-CRM
6, rue Boussingault
F-67083 Strasbourg Cedex, France

Progress in Colloid & Polymer Science   Progr Colloid Polym Sci 91:109–112 (1993)

Dynamic scattering from ternary mixtures of polymers in solution

M. Benmouna[1]), E. W. Fischer[2]), H. Benoit[3]), Z. Benmansour[1]), and T. A. Vilgis[2])

[1]) University of Tlemcen, Institute of Sciences, Physics Department, Tlemcen, Algeria
[2]) Max-Planck-Institut für Polymerforschung, Mainz, FRG
[3]) Institut Charles Sadron, Strasbourg, France

Abstract: The dynamic scattering properties of ternary mixtures made up of two polymer components and a solvent are considered. Examples of symmetrical and nonsymmetrical mixtures are examined. Particular attention is focused on the interpretation of the eigenmodes. It was recently shown that the eigenmodes for ternary mixtures could be rigorously identified with the cooperative and interdiffusive modes only if the two polymer components are identical, except for their optical properties. Here, we consider several mixtures made up of different polymer constituents and compare the variations of the relaxation frequencies with the scattering wave vector q. In all the examples considered, the behavior of (Γ, Γ_T) and (Γ', Γ_I) are found to be qualitatively the same. This leads to the conclusion that, although the eigenmodes can be identified rigorously as the cooperative and interdiffusive dynamical processes only for symmetrical mixtures with a finite χ-parameter, they can still be so-identified approximately even in the case of different polymers and nonsymmetrical mixtures under certain experimental conditions.

Key words: Dynamic scattering – ternary mixtures – cooperative mode – interdiffusive mode

Introduction

Systematic quasi-elastic-light scattering studies of several mixtures made up of two polymers of different nature and a solvent were reported recently by Borsali et al. [1–3]. These authors investigated the variations of the frequencies and amplitudes of the eigenmodes with parameters such as the concentration, composition, etc. The data were interpreted successfully using a simple theoretical model based on the random-phase approximation (RPA) for the static properties and the Rouse limit for the hydrodynamic properties. Similar methods have also been used by other authors to analyze data of ternary mixtures obtained by QELS [4–8] and neutron spin echo (NSE) [9] techniques. The eigenmodes with frequencies Γ and Γ' are identified as the cooperative and interdiffusive modes even in the conditions where the two polymers are different. Recently, it has been shown that these eigenmodes can be rigorously interpreted as

such only in the special case where the two polymers are identical except for their contrast factors or $(\partial n/\partial c)$. This was shown by Akcasu et al. [10, 11] from the resolution of the eigenvalue problem for the equation of motion of the particle concentrations where the first cumulant matrix plays the role of the dynamical operator. A similar analysis was made some time ago by Pusey et al. [12] and Davies et al. [13]. The former authors developed the method and applied it to mixtures of hard spheres and the latter were interested more in its application to bimodal dextran systems in aqueous solutions.

In this paper we consider a few examples of symmetrical and nonsymmetrical systems and discuss the variations of several relaxation frequencies with the wave vector q under certain conditions of dissymmetry in size, thermodynamic properties and other relevant parameters. We also briefly discuss the case of weakly charged polymers. The theoretical formalism was discussed in detail and in several papers [10, 14, 15]. Here, we present a few applications in an attempt to shed light on the interpretation of the eigenmodes for symmetrical and nonsymmetrical mixtures.

Symmetrical mixtures

Let us first consider the case of a symmetrical mixture of two homopolymers having the same form factors $P_1(q) = P_2(q) = P(q)$, at a composition $x = 1/2$, and satisfying the thermodynamic relationship between excluded-volume parameters $v_{11} = v_{22} = v_{12} - \chi = v$. There are only two partial structure factors, $S_{11}(q) = S_{22}(q) = S(q)$ and $S_{12}(q) = S'(q)$, which are easily obtained as

$$S(q) = \frac{\phi NP(1 + v\phi NP/2)}{2D}, \qquad (1)$$

$$S'(q) = -\frac{(v + \chi)\phi^2 N^2 P^2}{4D}, \qquad (2)$$

$$D = \left(1 - \frac{\chi}{2}\phi NP\right)\left[1 + \left(v + \frac{\chi}{2}\right)\phi NP\right]. \qquad (3)$$

In this case the eigenmodes are identified rigorously with the cooperative and interdiffusive modes and, therefore, we have $\Gamma \equiv \Gamma_T$ and $\Gamma' \equiv \Gamma_I$. We shall display the results in the Rouse limit, where the frequencies and the amplitudes can be given simultaneously and written as follows:

$$\frac{(\Gamma = \Gamma_T)}{q^2 D_0} = \frac{\phi N}{S_T(q)} = \frac{1}{P(q)} + \left(v + \frac{\chi}{2}\right)\phi N, \qquad (4)$$

$$\frac{(\Gamma' = \Gamma_I)}{q^2 D_0} = \frac{4\phi N}{S_I(q)} = \frac{1}{P(q)} - \frac{\chi}{2}\phi N. \qquad (5)$$

One must recall that the Rouse limit is valid only for concentrations above the overlap threshold ϕ^*, defined roughly by $v\phi^* N \approx 1$, where the hydrodynamic interaction is essentially screened. Equation (4) shows that Γ_T corresponds to the cooperative relaxation frequency for identical homopolymers with a total concentration ϕ and a slightly enhanced excluded-volume parameter $v + \chi/2$ due to the additional repulsion between unlike monomers. The interdiffusion mode displayed in Eq. (5) has a more interesting feature because it is similar to the result one would obtain in the bulk limit with a reduced interaction parameter $\chi\phi$. Furthermore, Eq. (5) shows that χ is

directly measurable either by static or dynamic scattering [16]. One has a reliable method in determining the variation of χ with the concentration from either static or dynamic scattering at least above ϕ^*, where both the RPA and the Rouse approximations are reasonable. Measurements using a similar method to the one proposed here were reported by Giebel et al. [3] using QELS on the system PDMS/PMMA/toluene.

Bimodal systems

These systems are made of two homopolymers of the same nature but with different molecular weights. The ratio of molecular weights is denoted by $y \equiv N_2/N_1$ and since monomers are identical, we have $v_{11} = v_{22} = v_{12} = v$. The eigenfrequencies Γ and Γ' can be expressed, in general, in terms of S_{ij} in the Rouse limit as follows:

$$\frac{(\Gamma, \Gamma')}{q^2 D_{01}} = \frac{\phi N_1}{2(S_{11}S_{22} - S_{12}^2)}$$
$$\times \{(1 - x)S_{11} + xS_{22}$$
$$\pm \sqrt{[(1 - x)S_{11} - xS_{22}]^2 + 4x(1 - x)S_{12}^2}\}, \qquad (6)$$

where x represents the composition of polymer 1, i.e. $x = \phi_1/\phi$, and $\phi = \phi_1 + \phi_2$. The cooperative relaxation frequency can be written explicitly since the amplitude $S_T(q) = S_{11} + S_{22} + 2S_{12}$ reduces to a simple form:

$$\Gamma_T/q^2 D_{01} = \frac{\phi N_1}{S_T(q)}$$
$$= \frac{1}{xP_1 + y(1 - x)P_2} + v\phi N_1. \qquad (7)$$

The interdiffusive mode is characterized by the amplitude

$$S_I(q) = S_{11}/x^2 + S_{22}/(1 - x)^2$$
$$- 2S_{12}/x(1 - x) \qquad (8)$$

and the frequency

$$\Gamma_I/q^2 D_{01} = \phi N_1/x(1 - x)S_I(q). \qquad (9)$$

In these equations, D_{01} denotes the diffusion coefficient of chain 1, i.e. $D_{01} = kT/N_1\xi$. The pairs of frequencies $(\Gamma; \Gamma_T)$ on the one hand, and (Γ', Γ_I) on the other hand, look quite different and it is rather difficult to judge their behavior by a simple inspection of these equations. Plots of the variations with qR_g of these frequencies in similar conditions show that $\Gamma \approx \Gamma_T$ and $\Gamma' \approx \Gamma_I$ for several values of the ratio y and the concentration ϕ (see ref. [17]). It was observed that for bimodal mixtures, even with wide differences in molecular weights, one may still identify approximately the eigenmodes as the cooperative and interdiffusive processes. We have also examined the case of different polymers, where in addition to the asymmetry in sizes, there is an asymmetry in thermodynamic parameters. It was found that Γ and Γ_T present only a slight discrepancy in the small q range and for high concentrations, but Γ' and Γ_I are practically identical and independent of the concentration when χ is maintained relatively small.

Nonsymmetric diblock copolymer

A similar analysis can be repeated for copolymers. The case of a copolymer with a composition $x = N_2/N$ different from $1/2$ and $v_{11} = v_{22} = v_{12} - x = v$, $\xi_1 = \xi_2 = \xi$ was considered and one finds that the plots of Γ and Γ_T on the one hand, and Γ' and Γ_I on the other hand, are practically identical for several values of x ranging from 0.1 to 0.5 [17].

Weakly charged polymers

The generalization of this investigation to weakly charged mixtures will be briefly discussed, emphasizing the aspect related to the main theme of the present paper, which is the interpretation of the eigenmodes and the comparison of their frequencies with those of the cooperative and interdiffusive processes. We summarize the observations made in two cases of charge distribution [17].

Similarly charged polymers ($f_1 = f_2 = f; \varepsilon = +1$)

This is the case where the monomers in the two components carry charges of the same nature, their amplitude is fe and $\varepsilon = +1$ because their signs are the same. One finds that the results obtained for the neutral system [14] remain valid but the excluded volume parameter v should be replaced by an ef-

fective parameter $v_{\text{eff}}(q)$, which is the sum of v and a q-dependent term, to account for the long-range electrostatic repulsion:

$$v_{\text{eff}}(q) = v + \alpha(q) \, f^2 . \tag{10}$$

The thermodynamic interaction between the two species remains unchanged and characterized by the parameter χ. This means that Γ and Γ' are practically equal to Γ_T and Γ_I, respectively, and that the eigenmodes retain the same interpretation as in the neutral limit.

Oppositely charged polymers ($f_1 = f_2 = f; \varepsilon = -1$)

The charges carried by monomers of the two species have the same value but different signs. One also finds that the results in the neutral limit remain valid but v must be replaced by $v_{\text{eff}}(q)$ and χ by $\chi_{\text{eff}}(q)$:

$$\chi_{\text{eff}}(q) = \chi - 2\alpha(q) \, f^2. \tag{11}$$

Interestingly enough, the interpretation of the eigenmodes remains unchanged and the cooperative mode is not affected by the long-range electrostatic interaction. A subtle compensation between the excluded-volume repulsion of monomers within the same species and attraction between monomers of different species takes place. This can be seen by recalling that the expression of Γ_T contains the term $v + \chi/2 \equiv v_{\text{eff}}(q) + \chi_{\text{eff}}(q)/2$, such that the relaxation of the total concentration fluctuations remains exactly the same as in the neutral limit. The interdiffusive mode is, however, strongly modified by the long-range attraction between the two species. This can be understood by imagining that single chains of one species, when diffusing into regions populated by chains of the other species, feel strong attractions due to electrostatic forces. The variation of Γ_I with q is found to have a similar behavior as in ordinary polyelectrolyte systems.

Conclusions

In this work we have presented a few examples of nonsymmetrical mixtures and shown that the dynamical processes governing the normal modes can still be interpreted approximately, as the cooperative and the interdiffusive processes. In some cases one finds not only a qualitative similarity between the wave vector dependencies of the corresponding frequencies but also a quantitative agreement as well. Only in cases of extreme dissymmetry one finds a noticeable numerical discrepancy, especially between Γ and Γ_T. This means that although rigorously speaking Γ is not equal to Γ_T and Γ' is not equal to Γ_I, the fact that they have qualitatively the same behavior in various conditions implies that they describe similar dynamical processes. This justifies perhaps the success of the RPA formalism which was used in interpreting QELS data on several neutral systems. In these experiments two modes were identified and interpreted successfully as the cooperative and the interdiffusive modes [1–3], although it was not explicitly stated that this identification is only an approximation. It has been brought to our attention, however, that there are examples of nonsymmetrical mixtures in which such an identification breaks down [10, 18].

Acknowledgements

One of us (M.B.) thanks the Max-Planck-Institut für Polymerforschung (Mainz, FRG) for hospitality during the period when this work was accomplished.

References

1. Borsali R, Duval M, Benmouna M (1989) Macromolecules 22:816
2. Borsali R, Duval M, Benmouna M (1989) Polymer 30:610
3. Giebel L, Borsali R, Fischer EW, Meier G (1990) Macromolecules 23:4054
4. Tirrell M (1984) Rubber Chem Technol 57:523
5. Hanley B, Tirrell M, Lodge T (1985) Polymer Bulletin (Berlin) 14:137
6. Numasawa N, Kuwamoto K, Nose T (1986) Macromolecules 19:2593
7. Brown W, Zhou P (1989) Macromolecules 22:3508
8. Aven MR, Cohen C (1990) Macromolecules 23:476
9. Csiba T, Jannink G, Durand G, Papoular R, Lapp A, Auvray L, Boué F, Cotton JP, Borsali R (1991) J Phys II 1:381
10. Akcasu A. Z., Nägele G, Klein R (1991) Macromolecules 24:4414
11. Akcasu A. Z. (1992) In: Brown W (ed) Dynamic Light Scattering; The Method and Some Applications. Oxford University Press, Oxford
12. Pusey PN, Fijnaut MM, Vrij A (1982) J Chem Phys 77:4270
13. Davies P, Snook I, van Megen W, Preston BN, Compar WD (1984) Macromolecules 17:2376
14. Benmouna M, Benoit H, Duval M, Akcasu ZA (1987) Macromolecules 20:1107
15. a) Benmouna M, Vilgis TA (1991) Macromolecules 24:3866; (b) Benmouna M, Vilgis TA, Polymer Networks Blends, to be published
16. Benmouna M, Fischer EW, Ewen B, Duval M (1992) J Polym Sci Part B: Polym Phys, submitted
17. Benmouna M, Benmansour Z, Fischer EW, Benoit H, Vilgis TA (1992) J Polym Sci Part B: Polym Phys 30:733
18. Akcasu Z. A., private communication

Received December 16, 1991;
accepted May 5, 1992

Authors' address:

Prof. Mustapha Benmouna
Max-Planck-Institut für Polymerforschung
Postfach 31 48
D-W-6500 Mainz, FRG

Progress in Colloid & Polymer Science   Progr Colloid Polym Sci 91:113–116 (1993)

Dynamics in concentrated polymer solutions studied using dynamic light scattering

W. Brown

Institute of Physical Chemistry, University of Uppsala, Sweden

Abstract: Concentrated solutions (volume fraction 0.1–0.9) of polystyrene (PS) in toluene and of polybutylacrylate (PBA) in dioxane have been examined using dynamic light scattering (DLS) as a function of angle and temperature. With PS there is a pronounced maximum in the cooperative diffusion coefficient (D_c) at about $\Phi = 0.5$ as the glassy state is approached. By contrast, with PBA, which has a very low T_g, there is a crossover to a stronger concentration dependence for D_c, which is attributed to an increase in solvent mobility. In the PS/toluene system at $\Phi = 0.78$ a transition is noted between the diffusion mode and the main relaxational mode. At 17 °C q-independence is observed and at 50 °C q^2-dependence. At $\Phi = 0.89$, both relaxations are observed simultaneously in the DLS time window, the slower process being a structural relaxation.

Key words: Photon correlation – polystyrene – concentrated solutions – relaxation times

Introduction

We have recently investigated concentrated polymer solutions using dynamic light scattering (DLS) and polarized Rayleigh–Brillouin scattering [1–4]. The purpose was to extend measurements to the concentration region separating the semidilute regime (volume fraction $\Phi < 0.15$) and the bulk polymer so as to ultimately understand the complex transitional behavior. This note focuses on the DLS data while the Rayleigh–Brillouin experiments are reported elsewhere.

The dynamics of semidilute solutions, both as regards *theta* sovents and thermodynamically good solvents, have been experimentally studied in detail; see, for example, refs. [5–7]. A summary of the theoretical treatments has been given by de Gennes [8]. The major aspect that has not yet been subject to more than a cursory theoretical treatment is the broad distribution of slow, q-independent, modes which has been shown to contribute the major part of the spectral density in semidilute θ-systems. The latter are now known to be related to the viscoelastic properties of the solution. Thus, a recent comparison between the DLS relaxation spectrum and the dynamic mechanical spectrum obtained from oscillatory shear measurements on the same semidilute solutions of polystyrene in the viscous solvent DOP [6] showed close similarities. Wang [9] has shown that the concentration fluctuations relax by cooperative diffusion and the viscoelasticity of the solution. These are mixed in the semidilute polymer solution to a degree depending on the frequency and a coupling parameter β, which is proportional to the difference in partial specific volumes of polymer and solvent. Further work in our group is being directed to a more quantitative interpretation following this line.

At the other end of the concentration scale, there are studies on amorphous bulk polymers near and above the glass transition temperature (T_g) (e.g., ref. [10]). The time correlation function of the density fluctuations in such systems is dominated by the primary or α-relaxation which arises from segmental motions in the polymer chain. In contrast to the properties in solution, bulk properties are insensitive to the chain length and chain entanglements. These chain motions may be also studied using dynamic mechanical or dielectric measurements. We note here that Wang and Fischer [11] have demonstrated that the density time correlation function for the bulk polymer is closely coupled to the relaxation of the longitudinal compliance. Exact relationships were provided linking the time correlation function from density fluctuations to the spectrum of retardation times of the longitudinal compliance.

Little work has been done until recently, however, either theoretically or experimentally, as regards the concentration interval extending between semidilute solutions and the bulk polymer. Pioneering investigations by Fytas and coworkers have been made on polymers with low concentrations of an added solvent, i.e., plasticized polymer systems [12]. These authors were able to demonstrate contributions from both concentration fluctuations and density fluctuations within the time window of the DLS experiment at the same time in the system polycyclohexylmethacrylate (PCHMA)/DOP. The two processes display different temperature dependences with cooperative diffusion having the weaker variation with temperature. A renewed attack on this field has been stimulated by the recently published theory of Wang [9] showing how the light scattering spectrum from a polymer solution arises from a combination of the concentration and density (or pressure) fluctuations. As the concentration is increased towards the bulk, the density fluctuations dominate the spectrum.

We present here DLS results on two polymer systems covering the volume fraction range: $\Phi = 0.1$ to 0.9 at different temperatures. (a) polystyrene (PS) in toluene where T_g is close to the measurement range [3] and (b) polybutylacrylate (PBA) in dioxane [4] where T_g is far below the measurement range. Very different behavior is observed in these systems.

Polystyrene/toluene

Figure 1 shows decay-time distributions over the concentration range studied. Up to $\Phi = 0.67$, the lineshape is close to single exponential and a single cooperative diffusive mode with the characteristic decay rate Γ_c is observed (although, significantly, it is possible to observe a trace quantity of viscoelastic modes [7]). The concentration dependence of the osmotic modulus is given by a power law

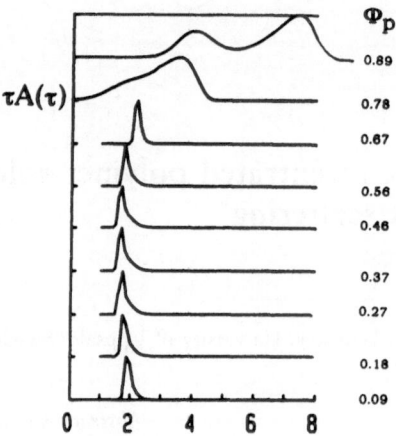

Fig. 1. Distribution of relaxation times for different polystyrene volume fractions (Φ) in toluene. The distributions are obtained by fitting to the sum of a Gaussian distribution and a generalized exponential (GEX) distribution (see, ref. [6]). An equal area plot is used: $\tau A(\tau)$ versus $\log \tau/\mu s$. Measurements correspond to $\Theta = 30°$ and $25°C$

with exponent 2.55 (blob theory [8] 2.25). The cooperative diffusion coefficient D_c shows an initial increase corresponding to a power law with an exponent 0.70, passing through a sharp maximum at about $\Phi = 0.5$, as was previously demonstrated for the PS/cyclohexane system (1); thereafter, D_c decreases very sharply. Such behavior has earlier been observed, for example, by Rehage and coworkers [13] and Patterson [14].

At $\Phi = 0.78$ a transition is observed between the cooperative diffusion mode and a dominant slow relaxational mode depending on the temperature of the measurements. The switchover from a situation dominated by concentration fluctuations to one apparently dominated by density fluctuations occurs over a surprisingly narrow interval in concentration and temperature and must reflect the close proximity to the glass–rubber transition. The two processes overlap and give rise to a mixed mode. This complex behavior is illustrated in Fig. 2 showing that q-independence is found at $17°C$ and q^2-dependence at $50°C$, while temperatures in between have an intermediate q-dependence.

At a concentration of $\Phi = 0.89$ the lineshape is bimodal, as is shown by the correlation function illustrated in Fig. 3. For this reason we have chosen to fit to the sum of a Gaussian distribution and a generalized exponential (GEX) distribution (for details see, ref. [6]). A stretched exponential form

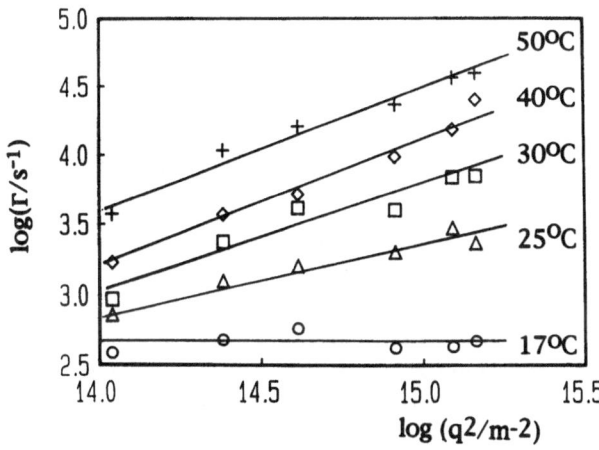

Fig. 2. Dependence of the relaxation rate on scattering vector (q^2) in a log–log diagram. Polystyrene/toluene system; volume fraction $\Phi = 0.78$ at the temperatures shown

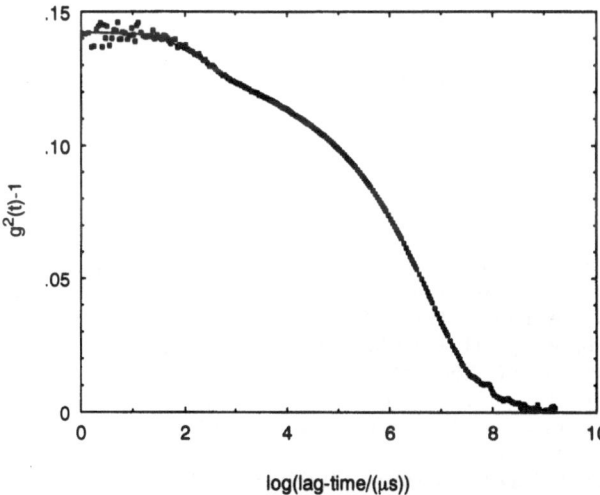

Fig. 3. Intensity–intensity (I_{VV}) time correlation function for polystyrene in toluene ($\Phi = 0.89$) at $\Theta = 60°$ and $30°C$

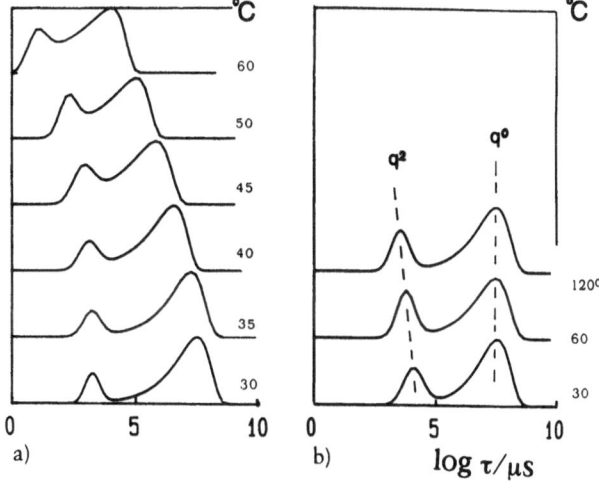

Fig. 4. a. Decay time distributions at different temperatures for the PS/toluene system; $\Phi = 0.89$. b) Distributions for different angles for solution with $\Phi = 0.89$

has no relevance for a distribution of this complexity. Figure 4a shows decay time distributions at different temperatures for the solution of $\Phi = 0.89$ and Fig. 4b similar distributions obtained at different angles on the same solution. Estimates of T_g were made using DSC and gave 275 K and 313 K for $\Phi = 0.78$ and 0.89, respectively. The faster process (Γ_c) observed at $\Phi = 0.89$ is attributed to concentration fluctuations probably associated with diffusion of the solvent. The much slower, q-independent process (Γ_d), corresponds to the slow structural relaxation of the polymer matrix. We note that Fytas et al. [12] have also observed concentration and density fluctuations simulta-

neously in the PCHMA/DOP system in the range 90 °C–125 °C, but here the fast mode corresponds to the density fluctuations and the slow mode to concentration fluctuations for the polymer. The coupling model of Ngai and coworkers (e.g., ref. [15]) was applied with considerable success in providing a theoretical framework. Typical of the density fluctuations reflecting primary segmental relaxation (in a polymer melt) is that the distribution of structural relaxation times is very broad, spanning over more than 10 decades (10^{-9}–10^3 s). Also characteristic of density fluctuations is a much stronger temperature dependence of the relaxation rate than is the case for the concentration fluctuations.

Polybutylacrylate/dioxane

In contrast to the PS/toluene system, PBA has a very low T_g-value (in the vicinity of 218 K). Thus, the large-scale motions of the chains would not be inhibited in the same way by the decrease in free-volume accompanying the glass transition. At 50 °C the lineshape approximates a single exponential and only a single cooperative mode can be extracted. It should be noted, however, that at lower temperatures a structural relaxation comes into the time window in the intermediate concentration range (0.4–0.6 g ml^{-1}) [19] and this is most likely attributed to viscoelastic relaxation as described by Wang [9].

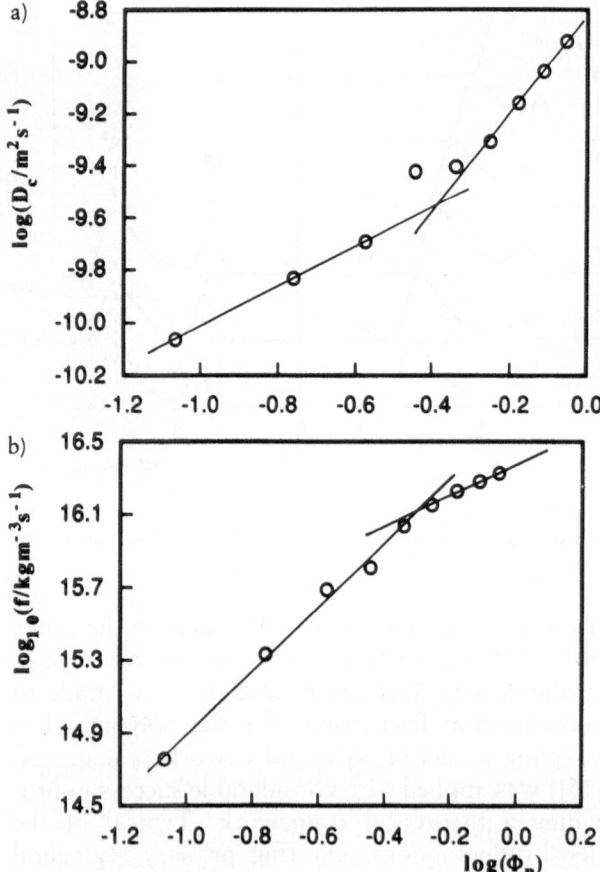

Fig. 5. Logarithmic dependence of (a) the cooperative diffusion coefficient and (b) the frictional coefficient at 50 °C on polymer volume fraction in the PBA/dioxane system

Figure 5a shows D_c as a function of Φ. The power law in the low Φ-region corresponds to $D_c \sim \Phi^{0.76}$ (for $\Phi < 0.3$), in excellent agreement with blob theory [8]. Contrary to the PS/toluene system, a maximum in D_c is not observed. Instead a crossover to a stronger concentration dependence is observed with $D_c \sim \Phi^{1.87}$ for $\Phi > 0.56$. The power-law exponent for the osmotic modulus determined by intensity light scattering is 2.68.

Figure 5b shows the concentration dependence of the friction coefficient, f, derived from the relationship: $D_c = M_{os}/f$. In the lower region the exponent is 1.76 (compared with the theoretical value of 1.5 from blob theory) and the exponent is 0.84 at the highest concentrations. By contrast, in the PS/toluene system, f increases strongly with Φ as the system approaches the glassy state and f is then presumably determined by a pronounced decrease in free volume. These changes in slope for D_c and

f in the PBA/dioxane system are interesting in light of recent observations by Lodge et al. [17, 18] on the solvent mobility in PS and polybutadiene (PBD) solutions. Here, PBD also has a very low T_g-value and exhibits properties close to those of PBA). In contrast to PS solutions, the solvent mobility in PBD solutions was found to increase on adding polymer; (i.e., this is consistent with an increase in free volume and a decrease in the effective solvent viscosity). We conclude that the observed effects on D_c and f are probably related to changes in solvent structure mediated by the polymer and which results in a decrease in the local solvent viscosity.

References

1. Brown W, Štěpánek P (1991) Macromolecules 24:5484
2. Brown W, Schillén K, Johnsen R.M, Koňák C, Dvoranek L (1992) Macromolecules 25:802
3. Brown W, Johnsen RM, Koňák C, Dvoranek L (1991) J Chem Phys 95:8568
4. Brown W, Johnsen RM, Koňák C, Dvoranek L (1992) J Chem Phys 96:6274
5. Nicolai T, Brown W, Johnsen RM, Štěpánek P (1990) Macromolecules 23:1165
6. Nicolai T, Brown W, Hvidt S, Heller K (1990) Macromolecules 23:5088
7. Brown W, Nicolai T (1990) Colloid Polym Sci 268:977
8. de Gennes P-G (1979) Scaling Concepts in Polymer Physics. Cornell University Press, London
9a. Wang CH (1991) J Chem Phys 95:3788
9b. Wang CH (1992) Macromolecules 25:1524
10. Fytas G (1989) Macromolecules 22:211
11. Wang CH, Fischer EW (1985) J Chem Phys 82:632, 4332
12. Fytas G, Floudas G, Ngai KL (1990) Macromolecules 23:1104
13. Rehage G, Fuhrmann J (1970) Discuss Faraday Soc 49:208
14. Patterson (1981) 27th Int Symp Macromol II:709
15. Ngai KL, Wang CH, Fytas G, Plazek DL, Plazek DJ (1987) J Chem Phys 86:4768
16. Lempert W, Wang CH (1981) Polym Prepr (Am Chem Soc Div Polym Chem) 22:80
17. Morris RL, Amelar S, Lodge TP (1988) J Chem Phys 89:6523
18. von Meerwall ED, Amelar S, Smeltzly MA, Lodge TP (1989) Macromolecules 22:295
19. Fytas G, personal communication

Received January 13, 1992;
accepted June 22, 1992

Author's address:

Wyn Brown
Inst. of Physical Chemistry
Univ. of Uppsala
Box 532
S-751 21 Uppsala, Sweden

Progress in Colloid & Polymer Science   Progr Colloid Polym Sci 91:117–120 (1993)

Dynamics of PS–PMMA diblock copolymers in toluene

M. Duval, H. Haida, J. P. Lingelser, and Y. Gallot

Institut Charles Sadron (CRM-EAHP) CNRS-ULP, Strasbourg, France

Abstract: The dynamic behaviour of two polystyrene–polymethylmethacrylate (PS–PMMA) diblock copolymers has been studied in toluene by quasielastic light scattering in the vicinity and above the overlap concentration C^* of the copolymer chains. At the macromolecular scale two relaxation modes have been found near C^*. The slow mode has been ascribed to the classical cooperative motion of the transient physical network formed by the copolymer molecules. It is identical to the motion that is observed in the equivalent binary PS/toluene solution where PS has the same molecular weight as the molecular weight of the whole copolymer and where the PS and the PS–PMMA concentrations are equal. The fast mode is a structural mode that reflects the fluctuations of concentration of one species with respect to the other. Above C^* a third motion appears that is not anticipated by the theories. The frequency of this mode is very low and decreases when the copolymer concentration increases. This mode is attributed to the formation of aggregates that can occur even at relatively low concentrations ($4C^*$) for the systems under investigation.

Key words: Diblock copolymer – dynamic light scattering – semidilute solution – nonselective solvent

Introduction

Theoretical investigations on the dynamic behaviour of A–B diblock copolymers in nonselective solvents have shown that such systems were characterized by two normal modes at the macromolecular scale [1]. When the two constituents of the copolymer differ only by their contrast factor these modes can be assimilated to the cooperative mode (Γ_1) of the transient physical network formed by the copolymer chains and to the interdiffusive mode (Γ_2) of one species with respect to the other [2]. The frequency Γ_2 of the latter mode greatly depends on the structure and on the composition of the copolymer. The theoretical expressions of the frequencies Γ_1 and Γ_2 of the two modes have been given in the case of an A–B (50/50) diblock copolymer [1].

Hydrogenated–deuterated polystyrene diblock copolymers (PSH–PSD) have been studied by neutron spin echo in semidilute solutions of hydrogenated–deuterated benzene (BzH–BzD) [3, 4]. The

structural mode (Γ_2) predicted by the theory has been observed in the zero average contrast condition where the cooperative mode vanishes.

In this work we have studied, by quasielastic light scattering (QELS), the dynamic behaviour of two polystyrene–polymethylmethacrylate (PS–PMMA) diblock copolymers in toluene, which is a good solvent for the two blocks. The solvent has the same refractive index as the PMMA block and the two modes (Γ_1, Γ_2) should be detected simultaneously.

Experimental

The experimental correlation functions of the intensity of the light scattered by the copolymer solutions were systematically compared with those obtained on the equivalent binary homopolystyrene (PS)/toluene solutions (see Fig. 1). The molecular weight of PS was the same as the molecular weight of the whole copolymer. The PS and

PS–PMMA samples were synthesized by anionic polymerization [5]. The polymers were fractionated [6] and the characteristics of the fractions which were used are given in Table 1. The range of the polymer concentration investigated for this study lies between C^* and $14C^*$, where C^* is the overlap concentration of the chains.

The dynamic light scattering measurements were performed on a home-built spectrometer [7]. They were done at several angles in the range $0.17 \leq qR_G \leq 1$, where q is the scattering wavevector and R_G is the radius of gyration of the polymers. The CONTIN method [8] has been used for the correlation function profile analysis of $G(D)$, where D is the diffusion coefficient given by

$$D = \Gamma/2q^2 \ . \tag{1}$$

The distribution functions $G(D)$ were always multimodal for the copolymer solutions in the whole range of concentration investigated, while they were monomodal for the equivalent PS solutions. The solutions were stable over a period of several weeks while they were under investigation.

Results and discussion

A typical distribution function of the diffusion coefficient obtained on the LG4F2/toluene solution, at $C = 1.80\%$ and at a scattering angle $\theta = 40°$, is shown in Fig. 2. Three relaxation modes can be observed in the copolymer solutions. Meanwhile, the equivalent binary PS/toluene solution shows only one mode. Furthermore, under the same optical conditions as for the copolymer solutions, the intensity of the light scattered by the equivalent binary polymethylmethacrylate (PMMA) solutions was equal to the intensity of the light scattered by toluene. The correlation functions of the light scattered by these PMMA/toluene solutions were flat for the small ($1 \, \mu s$) and large ($60 \, \mu s$) delay times used in this study.

The frequency (Γ_s) of the slow mode varies linearly with q^2 and goes to zero at zero scattering angle. It decreases when the copolymer concentration increases (see Fig. 3a and b). This mode is a diffusive mode that is specific to the copolymer solutions. It cannot be compared with the slow mode observed in binary PS/solvent systems because, in this range of concentration, the PS/toluene solutions were monomodal. The amplitude of

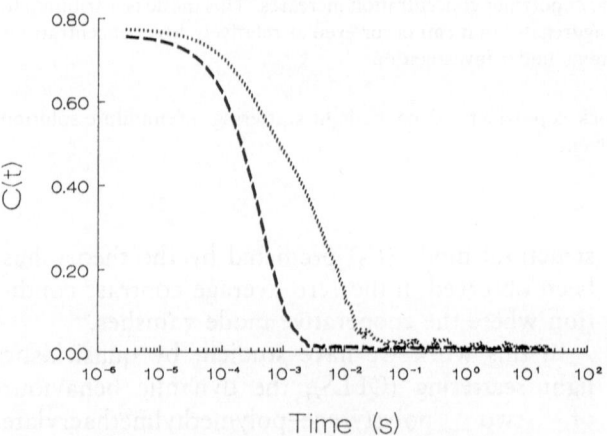

Fig 1. Experimental correlation functions measured at a scattering angle of $20°$ at $C = 4.2 \cdot 10^{-2} \, \text{g cm}^{-3}$. Multiple-tau correlogram: ($- - -$) PS531; ($\cdots$) LG3F3

Table 1. Polymer characteristics

Sample	$M_w^{a)}$	Weight fraction of PS$^{b)}$	$M_w/M_n^{c)}$
PS531	354 000		1.30
LG3F3	341 000	0.47	1.03
PSS52	630 000		1.17
LG4F2	640 000	0.38	1.06

$^{a)}$ Static light scattering
$^{b)}$ Determined via elemental analysis
$^{c)}$ Size exclusion chromatography in THF

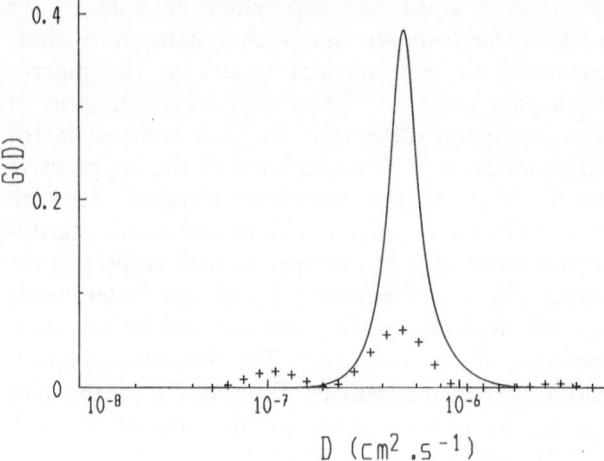

Fig. 2. Diffusion coefficient distribution $G(D)$ vs. D obtained from the CONTIN analysis: (———) PSS52: $C = 1.82 \cdot 10^{-2} \, \text{g cm}^{-3}$, $\theta = 20°$, $\Delta\tau = 3 \, \mu s$; (+) LG4F2: $C = 1.80 \cdot 10^{-2} \, \text{g cm}^{-3}$, $\theta = 40°$, $\Delta\tau = 6 \, \mu s$

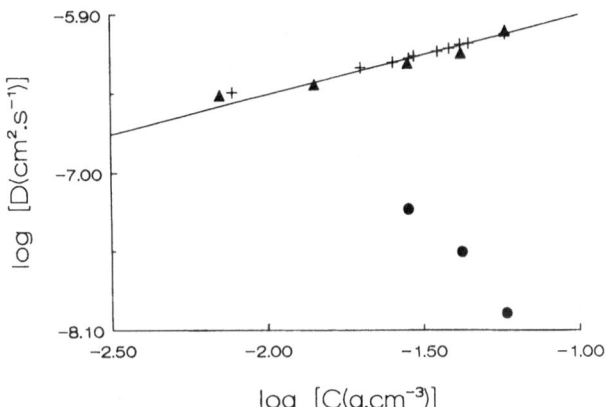

Fig. 3a. Variation of the diffusion coefficient as a function of polymer concentration: (+) PS531; (▲, ●) LG3F3 (intermediate and slow modes)

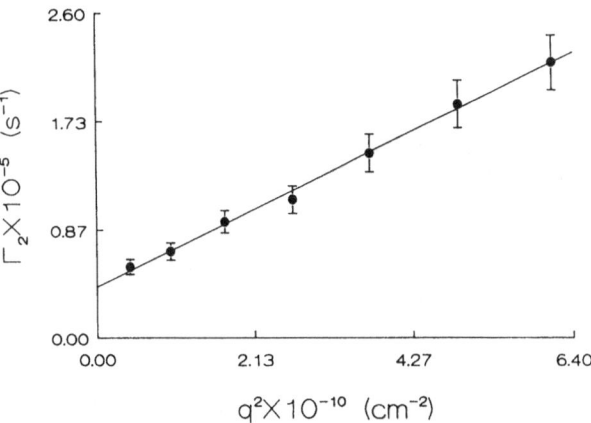

Fig. 4. Variation of the frequency of the fast mode as a function of the square of the scattering wavevector. Copolymer LG4F2; concentration: $0.67 \cdot 10^{-2}$ g cm^{-3}

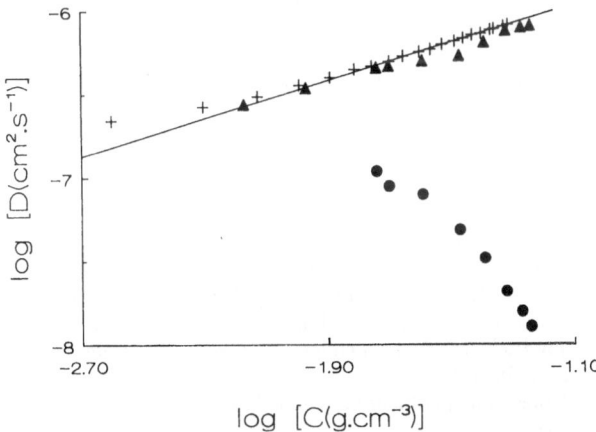

Fig. 3b. Variation of the diffusion coefficient as a function of polymer concentration: (+) PSS52; (▲, ●) LG4F2 (intermediate and slow modes)

this mode increases when the copolymer concentration increases. There is a threshold of concentration which depends on the molecular weight of the copolymer and below which this slow mode vanishes. Moreover, static light scattering measurements on these copolymer solutions have shown an increase of the scattered intensity at small angles. This slow mode has been attributed to the diffusion of the aggregates which are formed by the copolymer molecules even in a good solvent for the PS and PMMA blocks. Fredrickson and Leibler [9] have shown theoretically that well-organized structures should be observed in A–B diblock copolymer solutions even in a good solvent for the A and B blocks as soon as there is a small

incompatibility between the A and B species. Unfortunately, the static scattering measurements integrated all the effects of scattering by aggregates and by isolated molecules [10]. Furthermore, a better characterization of the copolymer solutions is difficult, due to the fact that the concentration and the molecular weight of the aggregates are unknown.

The frequency (Γ_1) of the intermediate mode observed in the semidilute copolymer solutions varies also linearly with q^2 and extrapolates to zero at zero scattering angle. This is a diffusive mode that is characterized by a diffusion coefficient which has the same value as the cooperative diffusion coefficient of the equivalent binary PS/toluene solution in the whole range of concentration investigated (see Fig. 3a and b). This mode has been assimilated to the slow normal mode predicted by the theory [1]. It is due to the fluctuations of concentration of the copolymer species with respect to the solvent. This is the cooperative mode of the network formed by the copolymer molecules.

The fast mode that appears in Fig. 2 in the copolymer solutions has a very small amplitude and cannot be studied quantitatively at high concentrations $(C \geq 4C^*)$ in the presence of the two other modes described previously. Nevertheless, the slow mode disappears at low polymer concentrations $(C < 3C^*$, see Fig. 3). This fast mode has been studied at two concentrations $(0.67 \cdot 10^{-2}$ and $1.06 \cdot 10^{-2}$ g/cm$^3)$ in the LG4F2/ toluene solutions. The variation of the frequency Γ_2 of this fast mode as a function of q^2 is represented in Fig. 4 for

$C = 0.67\%$. The frequency Γ_2 varies linearly with q^2 but tends to a constant different from zero at zero scattering angle. The same variation was obtained at $C = 1.06\%$. This mode of relaxation is a structural mode. It has been assimilated to the fast normal mode predicted by the theory [1]. It reflects the fluctuations of concentration of the S species with respect to the MMA species.

Recently, Konàk and Podesva [11] have studied a styrene–isoprene diblock copolymer in 1,1-diphenylethylene. Three relaxation modes were observed by these authors: the two modes of the theory and a slow diffusive mode due to the diffusion of aggregates in the copolymer solutions. Borsali et al. [12] have studied a PS–PMMA diblock copolymer in toluene. Two relaxation modes were found. The authors have interpreted these modes as the structural and diffusive modes predicted by the theory. We think that the fast mode has to be interpreted as the diffusive mode of the theory and that the slow mode should represent the diffusive motion of aggregates. A comprehensive discussion of the interpretation of these results has been published elsewhere [13].

The variation of the experimental ratio of the frequency of the fast mode to the frequency of the intermediate mode is drawn as a function of q^2 in Fig. 5. The theoretical curve which is drawn in the same figure has been calculated according to several approximations. The structure factors have been approximated by the Debye function. The quality of the solvent was supposed to be the same for both species S and MMA. The Flory interaction parameter χ between the S and MMA units has been estimated by an interpolation of experimental measurements on the PS/PMMA/toluene ternary system [14]. The experimental variation drawn in Fig. 5 is in good agreement with the theoretical variation where the hydrodynamical interactions [15] have been introduced.

References

1. Benmouna M, Benoit H, Duval M, Akcasu AZ (1987) Macromolecules 20:1107–1112
2. Akcasu AZ, Nägele G, Klein R (1991) Macromolecules 24:4408–4422
3. Borsali R, Benoit H, Legrand JF, Duval M, Picot C, Benmouna M, Farago B (1989) Macromolecules 22:4119–4121
4. Duval M, Picot C, Benoit H, Borsali R, Benmouna M, Lartigue C (1991) Macromolecules 24:3185–3188
5. Freyss D, Leng M, Rempp P (1964) Bull Soc Chim Fr 221–224
6. Haïda H, Lingelser JP, Gallot Y, Duval M (1991) Makromol Chem 192:2701–2711
7. Duval M, Coles HJ (1980) Rev Phys Appl 15:1399–1408
8. Provencher SW (1979) Makromol Chem 180:201–209
9. Fredrickson GH, Leibler L (1989) Macromolecules 22:1238–1250
10. Koberstein JT, Picot C, Benoit H (1985) Polymer 26:673–681
11. Konák C, Podesva J (1991) Macromolecules 24:6502–6504
12. Borsali R, Fisher EW, Benmouna M (1991) Phys Rev A 43:5732–5735
13. Duval M, Haïda H, Lingelser JP, Gallot Y (1991) Macromolecules 24:6867–6869
14. Ould-Kaddour L (1988) Thesis, University of Strasbourg
15. Yamakawa H (1971) Modern Theory of Polymer Solutions. Harper and Row, New York, pp 264–265

Received January 26, 1992;
accepted June 3, 1992

Authors' address:

Dr. Duval Michel
Institut Charles Sadron (CRM-EAHP)
6 rue Boussingault
67083 Strasbourg Cedex, France

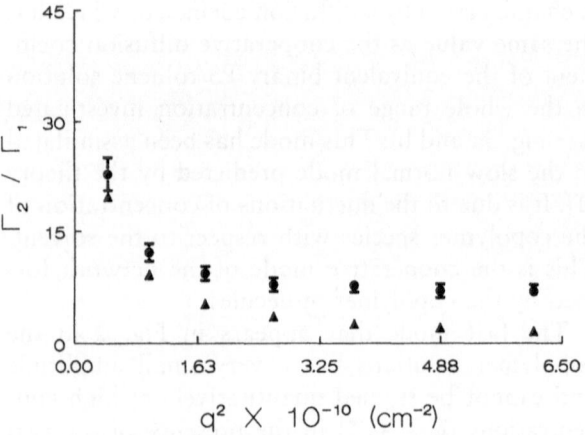

Fig. 5. Theoretical (▲) and experimental (●) variation of the ratio of the frequency of the fast mode to the intermediate mode as a function of the square of the scattering wavevector. Copolymer LG4F2; concentration: $0.67 \cdot 10^{-2}\,\mathrm{g\,cm^{-3}}$

Progress in Colloid & Polymer Science   Progr Colloid Polym Sci 91:121–123 (1993)

The effect of microscopic spatial restrictions on the segmental diffusion of dense polymer systems: Their observation and analysis by neutron spin echo spectroscopy

B. Ewen[1]), D. Richter[2]), B. Farago[3]), and U. Maschke[1])

[1]) Max-Planck-Institut für Polymerforschung, Mainz, FRG
[2]) Institut für Festkörperforschung, KFA Jülich, FRG
[3]) Institut Laue-Langevin, Grenoble, France

Abstract: The high-resolution neutron spin echo (NSE) spectroscopy provides the unique chance to study the segmental diffusion of polymeric systems simultaneously in space and time. In particular, information on the internal relaxation of single molecules or parts of single molecules may be obtained from dense systems, if the method of labelling by hydrogen deuterium exchange is used.

Whereas in general, a single Rouse dynamics is observed at short times, deviations towards a time-independent behaviour occur on larger time scales, if spatial constraints become effective. Provided that these deviations exhibit a systematic dependence on the magnitude of the scattering vector Q, they allow to derive a well-defined spatial regime where unrestricted Rouse dynamics is dominant.

In this contribution the effect of three completely different kinds of spatial constraints on the segmental diffusion occurring in high molecular weight homopolymer melts, in the compatible regime of low molecular weight polymer blends and in polymer networks, respectively, will be made evident and analysed.

Key words: Segmental diffusion – neutron spin echo spectroscopy – melts – polymer mixtures – networks

Introduction

The complex macroscopic dynamical behaviour of high molecular weight polymer systems in the liquid or rubber state obviously has to be related to the fact that segmental diffusion is by no means uniform on all intramolecular length and time scales [1]. This non-uniformity is attributed to the existence of constraints which act as spatial restrictions with respect to the segmental relaxation. However, up to very recently, the existence and the properties of these constraints were not directly accessible by microscopic experiments. This non-satisfying situation changed completely when quasi-elastic neutron scattering and, in particular, the neutron spin echo (NSE) spectroscopy was used to solve this problem. The power and uniqueness of this method results from the fact that it provides a direct simultaneous access to the microscopic motional behaviour in space and time, that single chain properties can be studied even in dense systems and that it is very sensitive to spatially restricted motional processes.

Theoretical

If spatial restrictions become dominant on the time scale of the experiment, the coherent dynamic structure factor $S(Q, t)/S(Q, 0)$ (Q magnitude of the scattering vector, t time) is composed of a time-independent contribution $S_0(Q)$ and a time-dependent part which decays from 1 to $S_0(Q)$ with increasing time. From the Q-dependence of $S_0(Q)$ information on the existence and properties of the constraints can be derived. This effect is known much better in the case of incoherent quasi-elastic neutron scattering, where the time-independent contribution is named elastic incoherent structure factor (EISF) [2]. There, its quantitative analysis is less difficult since only the spatial restriction of each individual scatterer has to be considered, whereas in the case of coherent scattering the spatial restrictions of all pairs of scatterers have to be taken into account.

The coherent analogue to the ESIF was calculated [3] for a high molecular weight polymer melt on the basis of the tube or reptation model [4], assuming that the Rouse dynamics [5] is stopped when the segments inside the tube experience the surrounding lateral constraints by which the fictitious tube is formed. Later on, Ronca [6] derived the complete coherent dynamic structure factor for the segmental diffusion in a polymer melt which includes spatially unrestricted (I) and spatially restricted Rouse dynamics (II) as well as the cross over between both regimes. In the time domain (I) $S(Q, t)/S(Q, 0)$ is a universal function of $Q^2 l^2 \sqrt{Wt}$. (l segment length, $W = 3kT/\xi l^2$, k Boltzmann constant, T temperature, ξ friction coefficient per segment), whereas in the time domain (II)

$$S(Q,t)/S(Q, 0) = \sqrt{\frac{\pi Q^2 d_t^2}{48}} \exp - \frac{Q^2 d_t^2}{24}$$

(d_t spatial extension of unconstrained Rouse dynamics or tube diameter) is independent of time.

Results and discussion

In addition to earlier NSE-investigations on a melt of alternating polyethylene polypropylene (PEP) copolymer [7, 8], where spatially restricted Rouse dynamics was observed unambiguously for the first time and a tube diameter $d_t = 48$ Å at $T = 492$ K was derived from the Ronca formula [6], the same effect with $d_t = 70$ Å at 473 K was also found in polydimethylsiloxane (PDMS). In order to see whether the terminal parts of the chain molecules experience the same constraints as the central part, similar measurements were performed on PDMS triblocks ($M_W \simeq 185.000$ g/mol), where either the central part (37%) or the terminal parts ($2 \times 40\%$) of the chains were labelled. Figure 1 shows the NSE spectra of the terminally labelled

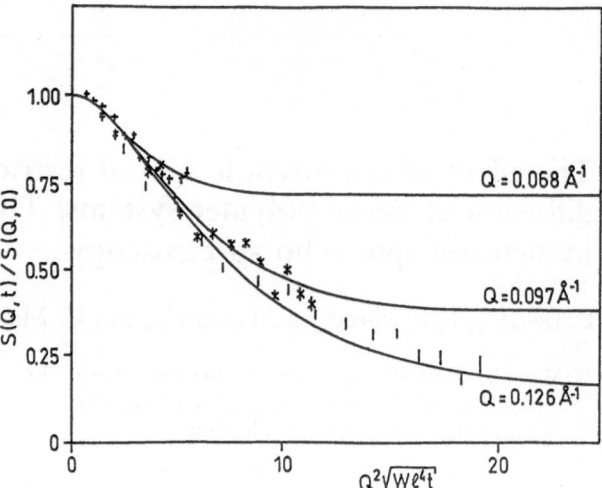

Fig. 1. Rouse scaling representation of the PDMS triblock data (terminal parts labelled ($0.068 \leq Q/\text{Å} \leq 0.126$) at $T = 473$ K. —— Fit with the Ronca approach

chains as a function of the Rouse variable $Q^2 l^2 \sqrt{Wt}$. At short times all data follow a master curve as expected for unrestricted Rouse dynamics whereas at longer times a Q dependent splitting to time-independent plateaus occurs. Within experimental error these findings are identical with the observations from completely as well as from centrally labelled PDMS chains. Thus, one can conclude that even at very large-tube diameters the segmental diffusion of the terminal chain parts is restricted in the same manner as that of the chain centre.

Spatially restricted segmental diffusion also becomes evident from NSE-investigations on the dynamics of junctions in fourfunctional PDMS model networks ($M_W^{mesh} \simeq 5500$ g/mol) [9]. The junctions fluctuate around their equilibrium positions. The spatial range which is covered by this motional process is rather extended. Its radius (24.5 Å) is nearly identical with the radius of gyration of the network strands. These findings fit nicely to the predictions of Flory's junction constraint model [10] and have to be considered as an important microscopic support of this model.

The segmental diffusion in a low molecular weight blend of deuterated PDMS and protonated polyethylmethylsiloxane (PEMS) at the critical concentration $\phi_{PEMS} = 0.49$ in the single-phase regime far away from the critical temperature T_c ($\Delta T = T - T_c \simeq 90\,^{\circ}\text{C}$) also shows strong deviations from unconstrained Rouse relaxation and

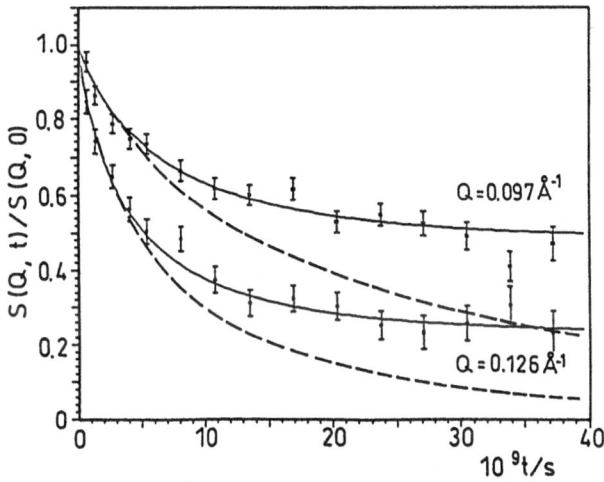

Fig. 2. Coherent dynamic structure factors $S(Q, t)/S(Q, 0)$ as a function of time for the mixture d-PDMS/PEMS ($\phi_{d\text{-PDMS}} \simeq 0.51$) at $T = 473$ K. $\times Q = 0.097 \text{ Å}^{-1}$; ● $Q = 0.125 \text{ Å}^{-1}$; – – – – fit of the short-time behaviour with the scattering law of the Rouse Model. —— fit with scattering law of the Ronca model with the Rouse rate of the short-time behaviour

the development of time-independent plateaus (see Fig. 2). There is no doubt that both pure species of the blend perform unconstrained Rouse dynamics. However, as long as the discrepancies with respect to the related relaxation rates W [11, 12] are not cleared away, it is still questionable whether spatially restricted Rouse relaxation occurs in the blend. A ratio of 10 between the segmental friction coefficients of both species as derived from the self-diffusion coefficients [11] is sufficient to produce spectra, as shown in Fig. 2, without taking into account spatial restrictions. If, on the other hand, both relaxation rates are nearly identical [12], the observed retardation of Rouse relaxation has to be attributed to the occurrence of spatial constraints. Since no molecular model is available for this situation, the Ronca formalism [6] was

used to estimate the length scale d of unconstrained Rouse dynamics. From this procedure $d = 59$ Å was derived, which exceeds the coil dimension ($< R_g^2 > 1/2 \simeq 40$ Å) only by a factor 1.5.

Moreover, the occurrence of spatial constraints far away from T_c does not seem to be unreasonable since the critical slowing down of the coefficient of interdiffusion also becomes first visible at comparable ΔT [12].

Acknowledgement

Two of the authors (B.E. and U.M.) gratefully acknowledge the financial support of the Bundesminister für Forschung und Technologie (BMFT), Bonn.

References

1. Graessley WW (1974) Adv Polym Sci 16:1
2. Springer T (1972) Quasielastic Neutron Scattering for Investigation of Diffusive Motions in Solids and Liquids. Springer Verlag, Berlin
3. de Gennes PG (1981) J Physique 42:735
4. Edwards SF (1967) Proc Phys Soc (London) 91:513
5. Rouse PE (1953) J Chem Phys 21:1272
6. Ronca GJ (1983) J Chem Phys 79:1031
7. Richter D, Farago B, Fetters LJ, Huang JS, Ewen B, Lartigue C (1990) Phys Rev Lett 64:1389
8. Richter D, Ewen B (1992) Suppl Coll & Polym Sci 270:
9. Oeser R, Ewen B, Richter D, Farago B (1988) Phys Rev Lett 60:1041
10. Flory JP (1977) J Chem Phys 66:5720
11. Fleischer G (1992) Suppl Coll & Polym Sci 270:
12. Momper B (1989) PhD Thesis, Mainz

Received February 14, 1992;
accepted May 10, 1992

Authors' address:

B. Ewen
Max-Planck-Institut für Polymerforschung
Ackermannweg 10
D-W-6500 Mainz, FRG

Polymer and solvent dynamics in a polystyrene/di-2-ethylhexyl phthalate solution

G. Floudas[1,2]), W. Steffen[2]), L. Giebel[2]), and G. Fytas[3])

[1]) Imperial College, Department of Chemical Engineering, London, UK
[2]) Max-Planck Institut für Polymerforschung, Mainz, FRG
[3]) Research Center of Crete, Greece

Abstract: Photon correlation spectroscopy (PCS) and dielectric spectroscopy (DS) are employed to study, respectively, the polymer and solvent dynamics in a polystyrene solution with di-2-ethylhexyl phthalate. The depolarized PCS and DS experiments revealed the presence of two primary relaxations in a PS/DOP solution displaying a broad but single calorimetric glass transition. The slow and fast primary relaxations are influenced mainly by the polymer and the solvent dynamics, respectively. Both relaxations are characterized by very broad distribution of relaxation times, at low *T*. This is discussed in terms of the environmental dissimilarities present in two component systems.

Key words: Photon correlation spectroscopy – dielectric relaxation – concentrated solutions

Introduction

In recent [1] dynamic light scattering study of concentrated polystyrene/di-2-ethylhexyl phthalate (PS/DOP) solutions two distinct components have been observed in the depolarized Rayleigh spectra (DRS), obtained with a Fabry–Perot interferometer. At high *T*, the broad interferometric component was due to "fast" reorienting solvent molecules, whereas the narrow component was due to the "slow" polymer dynamics. From the former component an effective solvent friction coefficient was extracted and compared with other polymer/solvent systems [2]. However, the latter spectral component, being instrumentally limited at high *T*, can only be resolved at lower *T*, near the glass transition temperature T_g, using PCS. It is the purpose of the present study to analyze the polymer and solvent dynamics in a PS/DOP solution with polystyrene composition $C_{PS} = 50\%$, at lower *T*, using depolarized PCS and DS.

Experimental

The weight-averaged molecular weight of PS in the solution with DOP was $3.2 \cdot 10^4$. The T_g and

ΔT_g for the solution with $C_{PS} = 50\%$ were 221 and 60 K, respectively, as determined by differential scanning calorimetry (DSC). For comparison, ΔT_g for bulk PS is only 7 K for the same heating rate (20 K min^{-1}). The time-correlation functions $G(t)$ of the depolarized light scattered intensity were measured under homodyne conditions at a scattering angle of 90° using the ALV-5000 multiple sampling time digital correlator. In homodyne detection, the desired normalized relaxation function is $C_{VH} = [G(t)/A - 1]^{0.5}$ with *A* being the baseline. The DS measurements were made with a Solartron–Schlumberger frequency analyzer FRA 1260 over the frequency range 10^{-1}–10^6 Hz. Details on the experimental procedure can be found elsewhere [1].

Results and discussion

The depolarized scattered intensity arises from fluctuations in the anisotropic part of the polarizability tensor and, thus, $C_{VH}(t)$ is given by

$$C_{VH}(t) = \langle \delta\alpha_{yz}(t) \cdot \delta\alpha_{yz}(0) \rangle , \qquad (1)$$

where $\delta\alpha_{yz} = \sum_j \alpha_{yz}(j, t) \exp\{i\mathbf{q} \cdot \mathbf{r}_j(t)\}$ denotes the

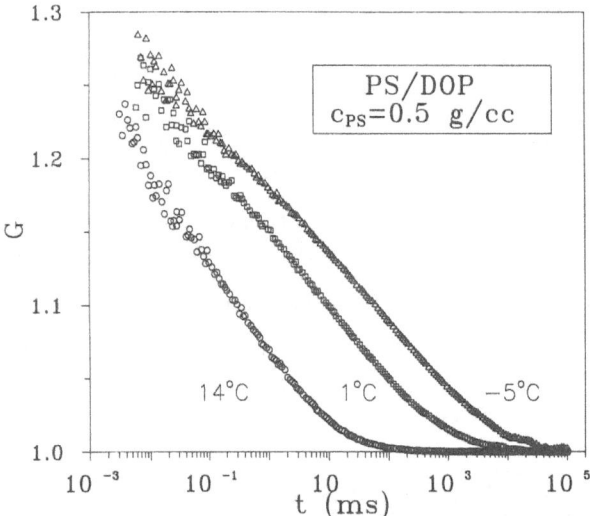

Fig. 1. Measured orientation correlation functions for the solution PS/DOP with $C_{PS} = 50\%$ at a scattering angle of 90°

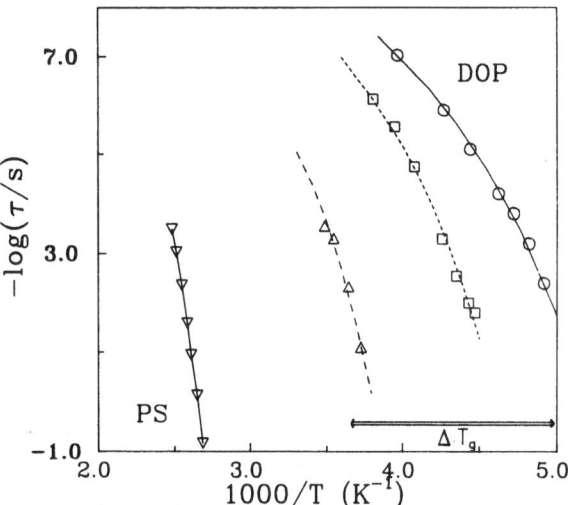

Fig. 2. Arrhenius plot of the relaxation times of the "slow" ($\triangle$) and "fast" $\square$) components present in the PS/DOP solution with $C_{PS} = 50\%$. The bulk polymer ($\triangledown$) and neat solvent ($\bigcirc$) relaxation times are shown for comparison. The broad glass transition temperature range ΔT_g (from DSC) is also indicated

spatial Fourier component of the polarizability density with $\alpha_{yz}(j, t)$ being the yz componet of the laboratory fixed polarizability tensor of the jth molecule located at $r_j(t)$. In Eq. (1) $q(= 4\pi n/\lambda \sin \Theta/2$ and n, λ and Θ are, respectively, the refractive index, wavelength and scattering angle) is parallel to the z-axis and the incident beam is polarized along the y-axis.

The correlation functions for orientation fluctuations in the PS/DOP solution with $C_{PS} = 50\%$ are shown in Fig. 1 as a function of temperature. The broad $C_{VH}(t)$ are dominated by two relaxation processes from which the slower (due to the polymer) can be now well analyzed. The orientation correlation functions $C_{VH}(t)$ in Fig. 1 can be described by the Kohlrausch–Williams–Watts (KWW) equation:

$$C_{VH}(t) = b \exp[- (t/\tau^*)^\beta], \quad 0 < \beta \leq 1, \quad (2)$$

where b, τ^* and β are fitting parameters describing, respectively, the contrast, dynamics and shape of the correlation function. The value of β for the "slow" $C_{VH}(t)$ component is 0.22 ± 0.02 indicating a very broad distribution of relaxation times. The relaxation times for the $C_{PS} = 50\%$ solution obtained from the Laplace inversion of the experimental $C_{VH}(t)$ are shown in Fig. 2 along with the times for bulk PS and neat DOP obtained from PCS and DS, respectively. The "fast" component – due to the solvent – in $C_{VH}(t)$ which relaxes outside the

correlator window can be analyzed using DS. DS selectively probes the solvent dynamics through the significant dipole moment of DOP. The DS relaxation times were obtained from the fit of the ε'' data to the Havriliak–Negami function followed by a Fourier transform to the time domain where a KWW function [Eq. (2)] was used. The DS times obtained with the procedure above are also shown in Fig. 2 for the 50% solution.

A pertinent feature of Fig. 2 is the existence of two non-Arrhenius relaxations in a PS/DOP solution having a single T_g. The "slow" and "fast" components, respectively are due to polymer and solvent relaxation. Evidently, the broad glass transition range ΔT_g also shown in Fig. 2, is a superposition of two distinct processes with very different dynamics. Therefore, phase separation occurs in the PS/DOP system in a scale which is much smaller than the probing length of the DSC (~ 100 Å). This is in accord with the small (10–20 Å) characteristic size of the cooperatively rearranging regions associated with the glass transition [3]. Furthermore, both components are characterized by broad distribution of relaxation times at low T ($\beta \approx 0.2$). At first glance, it seems difficult to rationalize the fact that the distribution of relaxation times for PS segments and DOP molecules in a PS/DOP solution is broader than in the bulk PS ($\beta_{PS} = 0.35$) and neat DOP ($\beta_{DOP} = 0.6$), respectively. This observation is related with recent experimental findings from two component systems: Polymer/additive mixtures [4], blends [5], symmetric and asymmetric block copolymers [6, 7] can produce very broad distributions as inferred from DSC, dynamic light scattering [4, 6, 7], dielectric [5] and mechanical experiments [8]. It seems, therefore, that it is not the nature (whether a polymer/additive mixture or a blend) not it is the chemical connectivity (blend vs. copolymer) which is responsible for the extremely broad distributions observed. This is probably the result of the environmental dissimilarities present in two component systems. It should, therefore, appear in any two component system irrespective of the nature and the connectivity of the two components.

References

1. Steffen W, Floudas G, Giebel L, Fytas G, in preparation
2. Floudas G, Fytas G, Brown W (1992) J Chem Phys 96:2164
3. Donth E (1991) J Non-Cryst Solids 131–133:204
4. Floudas G, Fytas G (1991) J Non-Cryst Solids 131–133:579
5. Zetsche A, Kremer F, Jung W, Schulze H (1990) Polymer 31:1883
6. Kanetakis J, Fytas G (1991) J Non-Cryst Solids 131–133:823
7. Gerharz B, Fischer EW, Fytas G (1991) Polymer Commun 32:469
8. Roland CM, Ngai KL (1991) Macromolecules 24:2261

Received January 21, 1992;
accepted April 30, 1992

Authors' address:

G. Floudas
Max-Planck Institut für Polymerforschung
D-W-6500, Mainz, FRG

Dynamic light scattering investigations on semidilute solutions of branched polyethylene

M. Helmstedt

Fachbereich Physik, Universität Leipzig, FRG

Abstract: Fractions of branched polyethylene were investigated by dynamic light scattering in 1,2,4-trichlorobenzene at 135 °C. The individual macromolecules and the microgel were separated by Laplace inversion of the autocorrelation functions. For branched polyethylene with $\sigma = 0.919$ g/cm^3 the relationship

$$D_0 = 2.437 \cdot 10^{-4} M_w^{-0.533}$$

is valid. In semidilute solutions two further modes were observed. The diffusion coefficient of the first mode does not depend on the angle of observation, and the slope of the concentration dependence is -1.90.

The diffusion coefficient of the second mode depends on the angle of observation. The slope of the concentration dependence is -2.83. It seems to be a diffusion of clusters.

Key words: Polyethylene – branching – polymer characterization – light scattering – semidilute solutions – polymer dynamics

Because polyethylene is soluble only at temperatures higher than 80°–135 °C, the preparation of solutions requires the exclusion of oxygen, higher temperatures of dissolution and filtration at these temperatures. This may be the main reason that only few results of dynamic light scattering measurements on polyethylene solutions have been published in the literature [1, 2, 3].

Diffusion and molecular parameters of fractions of branched polyethylene were investigated by static and dynamic light scattering in 1,2,4-trichlorobenzene at 135 °C with a home-built scattering photometer [4] and a modified SOFICA apparatus, respectively. A publication of the results on dilute solutions of fractions with narrow distributions of the molar masses is in preparation.

Here, light scattering measurements on fractions are described, which were prepared by precipitation fractionation [5]. The scattering behaviour of dilute and semidilute solutions, especially at small angles, is strongly influenced by the microgel content. Some percent of highly branched microgel completely "masks" the main part of fraction I in the Zimm plot of the measurement of dilute solutions of this fraction (Fig. 1). The results are given in the legend of the figure and reflect the microgel part of the sample only. The scattering intensity is the sum of the intensities of the light scattered by dissolved macromolecules and microgel particles. The components can be separated by a two-component fit of the curve for the angular dependency at infinite dilution.

In dynamic light scattering the components can be separated by Laplace inversion of the electric field autocorrelation functions by Provencher's CONTIN or "non-negative least-squares" procedures [6, 7]. An example for the component separation of an autocorrelation function, measured on a dilute solution of sample I, is given in Fig. 2. For the individual molecules the observation of inner modes of motion was excluded by special experimental conditions, that means measurements at low angles of observation between 30° and 60°.

Molecular characterization is the basis for the study of the behaviour of branched macromolecules in semidilute solutions by dynamic light scattering

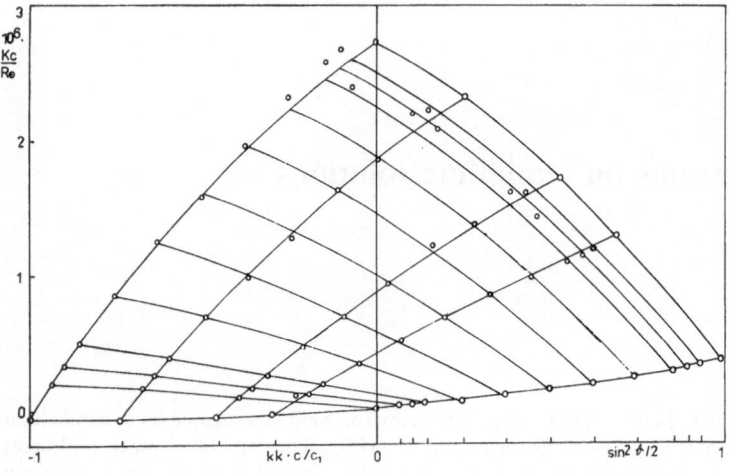

Fig. 1. Zimm plot of the branched poly-ethylene fraction I with $M_w(GPC) = 503 \cdot 10^3$ g/mol and $M_w/M_n = 3.2$ [5] in 1,2,4-trichlorobenzene at 135 °C; $M_{w,app}(LS) = 1.79 \cdot 10^7$ g/mol, $R_G = 85.7$ nm

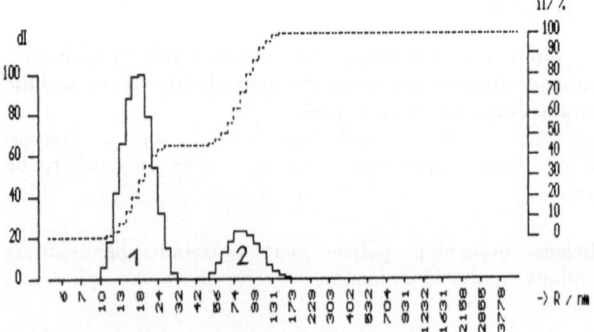

Fig. 2. Integral (*il*) and differential (*dl*) scattering intensity (Laplace inversion of the field autocorrelation function) of fraction I in 1,2,4-trichlorobenzene at 135 °C ($c = 8.85 \cdot 10^{-3}$ g/cm^3, $\vartheta = 30°$, $\Delta t = 0.05$ ms). Component 1: macromolecules $R_H = 19.4$ nm; component 2: microgel $R_H = 99.4$ nm

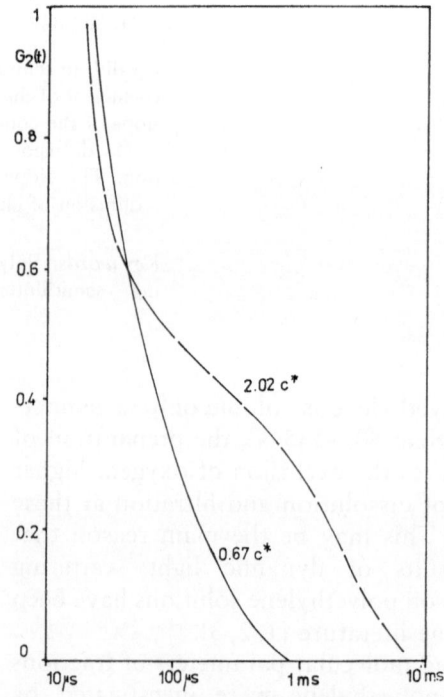

Fig. 3. Intensity correlation functions of solutions of fraction II of branched polyethylene in 1,2,4-trichlorobenzene at 135 °C ($M_w(GPC) = 303 \cdot 10^3$ g/mol, $M_w/M_n = 1.78$, $c^* = 6.1 \cdot 10^{-3}$ g/cm^3)

scattering, which is of interest in the field of dynamics of non-linear polymers in solution. At increasing concentrations the diffusion of macro-molecules becomes more and more complex. In Fig. 3 correlation functions of solutions with concentrations below and above the overlap concentration are shown ($c^* = 3M/4\pi R_G^3 N_A \approx 1/[\eta]$ overlap concentration, N_A is the Avogadro number).

In Fig. 4 are plotted the translational diffusion coefficients of the components, separated by NNLS, over a broad concentration range. At a concentration $c < 2c^*$ we observe the translational diffusion of the individual macromolecules (I) and the microgel (II, $D_0 \approx 2 \cdot 10^{-8}$ cm^2/s). The broken line at II is another, more speculative, possibility to interpret the data.

According to the relationship for branched poly-ethylene with density $\sigma = 0.919$ g/cm^3 (with $D_0 = D_t$ at $c = 0$) [8],

$$D_0 = 2.437 \cdot 10^{-4} \, M_w^{-0.533} \, ,$$

the translational diffusion coefficient for the individual macromolecules with $M_w = 503 \cdot 10^3$ g/mol

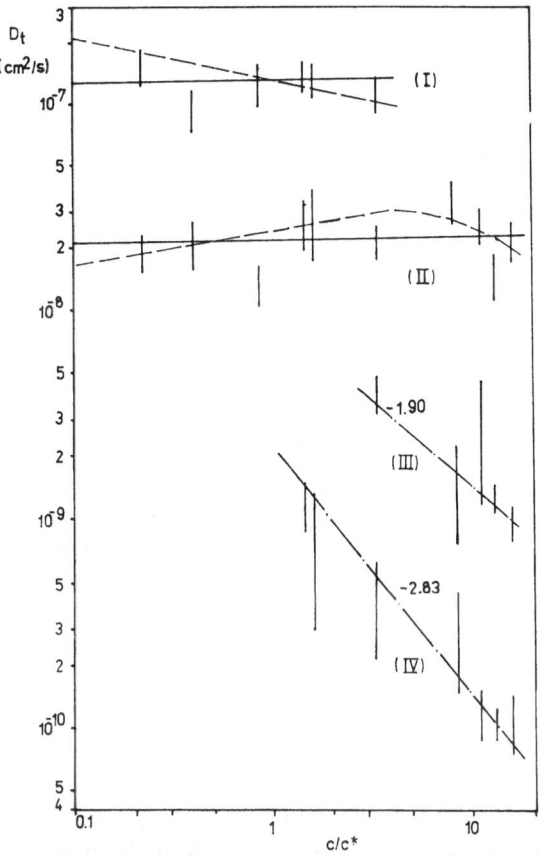

Fig. 4. Diffusion coefficients for semidilute solutions of fraction I in 1,2,4-trichlorobenzene at 135 °C, separated by the NNLS procedure. The numbers are the exponents a in $D_{slow}/D_0 = (c/c^*)^a$

is $D_0 = 2.2 \cdot 10^{-7}$ cm^2/s, according to the broken line I in Fig. 4, which is a guide for the eye only.

The diffusion of individual macromolecules disappears in the concentration range $c > 2c^*$ and two further processes become more and more intensive.

The diffusion coefficients of the first mode with the slope -1.90 do not depend on the angle of observation (III). The nature of this mode is still unknown; probably, it is related to the cooperative motion of polymer chains rise to local concentration fluctuations.

In contrast with these findings the diffusion coefficient of the slowest process (IV) with the slope of the concentration dependence of -2.83 depends strongly on the observation angle. In this case we assume a diffusion of clusters of macromolecules. The clusters have a radius of some micrometres according to the diffusion coefficient D of $5 \cdot 10^{-9}$ cm^2/s.

The discussion of the effects should be repeated after further detailed measurements on narrow fractions of branched polyethylene.

Acknowledgements

The author is grateful to Dr. E. Brauer and Dr. H. Wiegleb (Leuna-Werke AG, Merseburg) for preparation of the polyethylene fractions and to Dr. O. Prochazka (Institute of Macromolecular Chemistry, Prague) for his permission to use the programm "Zimm" for the evaluation of the static light scattering measurements.

References

1. Chu B, Onclin M (1984) J Phys Chem 88:6566–6575
2. Pope JW, Chu B (1984) Macromolecules 17:2633–2640
3. Helmstedt M, Fleischer G, Roth HK (1982) Plaste Kautschuk 29:628–630
4. Helmstedt M (1988) Makromol Chem Makromol Symp 18:37–52
5. Brauer E, Wiegleb H, private communication
6. Morrison JD, Grabowski EF, Herb OA (1985) Langmuir 1:496–501
7. Schäfer H, Wünsche P, Phys Stat Solidi (A), in preparation
8. Helmstedt M, Polymer, in preparation

Received January 16, 1992;
accepted June 12, 1992

Author's address:

Dr. Martin Helmstedt
Fachbereich Physik
Universität Leipzig
Linnéstraße 5
D-O-7010 Leipzig, FRG

Progress in Colloid & Polymer Science   Progr Colloid Polym Sci 91:130–134 (1993)

On the dynamics of dense polymer systems

D. Richter[1]), B. Ewen[2]), L. J. Fetters[3]), J. S. Huang[3]) and B. Farago[4])

[1]) IFF des Forschungszentrums Jülich, FRG
[2]) MPI für Polymerforschung Mainz, FRG
[3]) Exxon Research and Engineering Co., Annandale NJ, USA
[4]) Institut Laue Langevin Grenoble, France

Abstract: We discuss experimental results on molecular motion in polymer melts, obtained by neutron spin-echo spectroscopy (NSE). For longer times we observe systematic deviations from the Rouse model, revealing the presence of a well-defined intermediate dynamical length scale beyond which density fluctuations within a given chain are strongly reduced. Its value is found to be in excellent agreement with the entanglement distance obtained from rheological measurements. Measurements of the temperature dependence and polymer volume fraction dependence of the entanglement distance give an insight into the molecular origin of entanglement constraints.

Key words: Reptation – entanglements – neutron spin-echo – polymer melts – microscopic dynamics

Introduction

High molecular weight polymeric liquids exhibit unusual dynamic properties: Depending upon the time scale of observation or temperature, the same polymer may respond elastically, showing rubber-like behavior or may flow like a liquid [1]. The rubber behavior which expresses itself by the so-called plateau regime in the dynamic modulus is commonly attributed to the effect of "entanglements". They are thought to stem from geometrical or topological constraints, mutually imposed by the interpenetrating chain molecules. Their molecular origin, however, is not well understood. The reptation theory of viscoelasticity [2, 3] is based on the further assumption that the geometrical constraints can be modeled by a tube confinement surrounding a given chain. At intermediate times the polymer dynamics are restricted to a curvilinear motion along the tube. The tube diameter d, thereby, may be interpreted as the distance between entanglements.

Theoretical models

If the chains could intersect freely, the chain dynamics would be described as thermal motion damped via a friction coefficient ζ. In this so-called Rouse model [4], the diffusing chain segment performs a random walk on the random chain profile. This convolution of two random processes leads to a mean square segment displacement $\langle r^2(t) \rangle \simeq l^2(Wt)^{1/2}$, with $W = 3kT/\zeta l^2$ being the Rouse rate [5] and l the segment length. The dynamic structure factor of such a Rouse chain scales with a universal "Rouse" variable $u = Q^2 l^2 \sqrt{Wt}$; the Q–t scaling results from the fact that besides the cutoff length scales R_g (size of the chain) and l, the model is self-similar.

The presence of a well-defined entanglement distance acting as an intermediate dynamic length scale changes the scaling behavior of $S(Q, t)$ and causes systematic Q-dependent deviations from the Rouse scaling. In the framework of reptation, De Gennes derived an explicit first-order expression for $S(Q, t)$ [6]:

$$S(Q, t)/S(Q, 0)$$

$$= 1 - \frac{Q^2 d^2}{36} + \frac{Q^2 d^2}{36} \exp\left(\frac{u^2}{36}\right) \operatorname{erfc}(u/6) . \quad (1)$$

$S(Q, t)$ describes the equilibrium of density fluctuations along the tube (local reptation), neglecting any decay due to Rouse models of a spatial extent smaller than the tube diameter. The important feature of Eq. (1) is the factor $Q \times d$, which introduces a new length scale. Due to the tube constraints, $S(Q, t)$ decays only partially to a certain Q-dependent fraction. The remaining "elastic" part is a consequence of long-living segment–segment correlations due to the tube confinement. For a quantitative analysis of scattering data originating from the crossover regime between short-time Rouse motion and local reptation, it is necessary to include the initial Rouse motion which was neglected by De Gennes. The only model in the literature providing an explicit expression for the dynamic structure factor in this regime is Ronca's effective medium model [7], using an ad hoc ansatz for the time-dependent friction.

The molecular origin of entanglements is not well understood but current thinking postulates them to originate mainly from the topological nature of long-chain molecules as being flexible, nearly one-dimensional uncrossable objects. The occurrence of entanglements is then governed by two length parameters, the step length of the Gaussian random walk $l_K = C_\infty l_0$ (C_∞ characteristic ratio, l_0 bond length) of the chains and the lateral distance between chains determining the amount of chain contour length per unit volume [8–13]. Scaling models and topological calculations are brought forward. Using the chain contour length density and the mutual uncrossability of chains as the determining features, Graessly and Edwards [8] developed a general scaling model on the basis of the experimentally found power law relationship between plateau modulus and polymer volume fraction $G_N^0 \propto \phi^a$. They find

$$d^2 \propto C_\infty^{4-a} \left(\frac{\rho\phi}{m_0}\right)^{1-a} l_0^{(5-3a)} , \qquad (2)$$

where ρ is the polymer density and m_0 the molecular weight/bond. More specific scaling models like the packing model [9] ($a = 3$) or different binary contact models [10–12] ($a = 2$; $7/3$) are realizations of Eq. (2) for particular values of a.

Topological calculations [13] go beyond scaling insofar as founded on the mathematical concept of topological invariants geometrical constraints are actually calculated and not conjectured on the basis of scaling arguments. Using Gaussian topological invariants, very recently Iwata and Edwards [13] introduced a new quantity, the topological interac-

tion parameter $\bar{\gamma}$, which measures the capability of a chain to entangle; $\bar{\gamma}$ is mainly determined by the diameter of a polymer chain. For the concentration dependence of the plateau modulus they calculate $G_N \propto \phi^a$, where a increases with decreasing concentration ($1.97 \leq a \leq 2.2$). For high concentrations their result agrees with $d \propto (\rho\phi/m_0)^{-1/2}$.

Experimentals and results

Given the still limited temporal resolution of neutron spin-echo ($t < 40$ ns), it was essential to select thermally stable linear polymers combining high flexibility, a large plateau modulus indicating important topological constraints, and a low monomeric friction coefficient yielding high segment mobility. For our experiments we chose polyisoprene (PI); the alternating copolymer of polyethylene propylene (PEP) which was obtained from hydrogenation of 1,4-polyisoprene and hydrogenated 1,4-polybutadiene (PEB-2), which essentially resembles polyethylene. The different polymer samples including their characteristics are listed in Table 1.

The quasielastic neutron scattering experiments were performed using the neutron spin-echo (NSE) spectrometer IN11 at the Institute Laue Langevin (ILL) in Grenoble [14]. The effect of entanglement constraints on the molecular dynamics in polymer melts has been investigated for all the three polymers [15]. The results for all the polymers are displayed in Table 2. Here we shall remark only on the data on PEP [16]. Figure 1 presents the measured dynamic structure factor for the PEP sample

Table 1. Polymer molecular characteristics

Sample	$M_w\,10^{-4\,a)}$	$M_z/M_w{}^{b)}$	$M_w/M_n{}^{b)}$
h_8-PI sample 1	5.7	1.02	1.03
d_8-PI sample 1	5.2	1.02	1.03
h_8-PI sample 2	7.40	1.02	1.04
d_8-PI sample 2	7.91	1.03	1.05
d_{10}-PEP	8.38	1.03	1.05
h_{10}-PEP	8.22	1.02	1.05
PEP triblock	8.78	1.03	1.06
h_8-PEB-2	7.05	1.04	1.04
d_8-PEB-2	7.32	1.02	1.04

a) Light scattering
b) Size exclusion chromatography

Table 2. NSE results on polymer melts

Sample	T (K)	NSE			Rheology	
		d (Å)	$10^9\zeta$ (dyn/cm)	τ_e (ns)	d (Å)	$10^9\zeta$ (dyn/cm)
PI 1	468		4.4 ± 0.3	—		1.2
PI 2	473	52 ± 1	3.8 ± 0.3	32	51 (298 K)	
PEP homopolymer	492	47.5 ± 0.4	3.1 ± 0.1	15	43.5 ± 2	1.9
PEP triblock	491	47.1 ± 0.7	2.4 ± 0.2	15		
PEB-2	509	$43.5 \pm 0.7^{a)}$	0.4 ± 0.04		35 (373 K) PE	0.3 (448 K)
		43.1 ± 0.9		5	42 (509) PEB-7	

$^{a)}$ Fit with local reptation

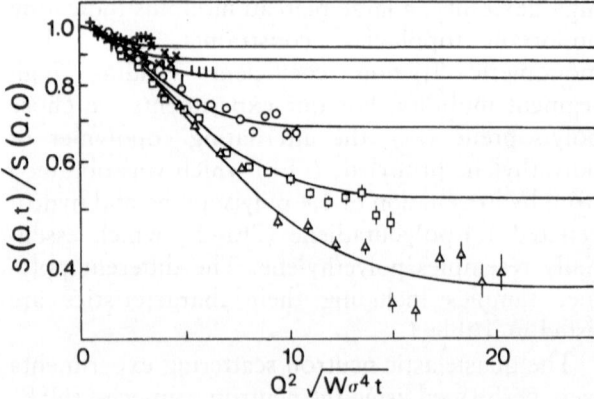

Fig. 1. Rouse scaling representation of the PEP data at 492 K: (+) $Q = 0.058$ Å^{-1}; (×) $Q = 0.068$ Å^{-1}; (|) $Q = 0.078$ Å^{-1}; (○) $Q = 0.097$ Å^{-1}; (□) $Q = 0.116$ Å^{-1}; (△) $Q = 0.135$ Å^{-1}. The solid lines are the result of a joint fit to the Ronca model

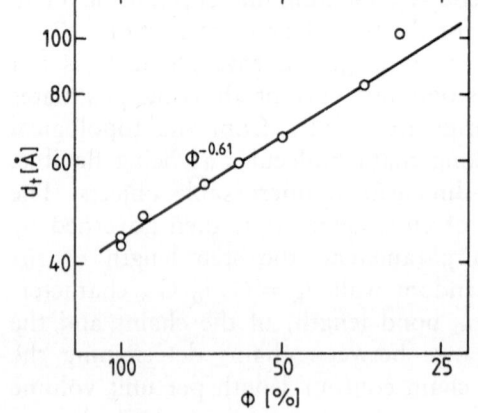

Fig. 2. Double logarithmic presentation of the entanglement distance or tube diameter d for PEB-2 at 509 K as a function of polymer volume fraction ϕ

at 492 K in a scaling form. The data are characterized by a common initial decay signifying the Rouse regime and a consecutive Q-dependent crossover into a plateau resulting from the presence of an intermediate dynamic length scale beyond which density fluctuations are strongly limited. The solid lines represent the result of a fit with the Ronca model. It allows a very satisfying description of the experimental data reproducing the line shape, the resulting sharp crossover, and the Q-dependence of the plateau levels. In order to compare with rheological results, we studied the plateau modulus of a high molecular weight PEP sample ($M_w = 170\,000$) at 500 °C using a Rheometrics System 4. The value obtained is 9.3×10^6 dyn/cm^2 resulting in a tube diameter or entanglement distance of 43 Å. This value com-

pares very well with the 47 Å measured by NSE on the microscopic scale.

In order to assess experimentally the different theories on entanglement formation, we varied systematically the two length parameters of importance. On PEB-2 we studied the dependence on contour length density while varying the polymer volume fraction at constant temperature ($T = 500$ K) by diluting with the oligomer $C_{19}D_{40}$ over a wide concentration range of $0.25 \leq \phi \leq 1$. The dependence on l_K was investigated by changing temperature and thereby C_∞ at a given ϕ. Figure 2 displays the resulting dependence of the entanglement distance on ϕ. The solid line corresponds to $d \propto \phi^{-0.61}$, describing up to $\phi = 0.35$ the experimental data very well. At $\phi = 0.3$, the d value is definitely larger than suggested by the $d \propto \phi^{-0.61}$

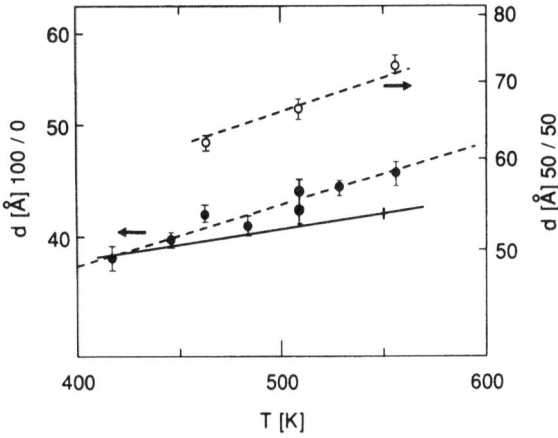

Fig. 3. Temperature dependence of the entanglement distance d of PEB-2: ($\bigcirc$, ϕ) = 1; ($\bullet$, ϕ) = 0.5. The dashed lines are guides to the eye. The solid line represents the scaling prediction (see text)

power law; at $\phi = 0.25$ an entanglement distance could not be determined any more with certainty.

We also addressed the temperature dependence of the microscopic dynamics of PEB-2. The results obtained for $\phi = 1.0$ and $\phi = 0.5$ are displayed in Fig. 3. They are compatible with an identical temperature coefficient for both concentrations. The solid line represents the best fit for the $\phi = 1$ data.

$$d = 23.5 \exp[(1.2 \pm 0.2) 10^{-3} T] \qquad (3)$$

Discussion

We now compare the predictions of the various scaling models with the results obtained from PEB-2. For the concentration dependence of the entanglement distance, we found $d \propto \phi^{-0.61}$. In terms of the general scaling model of Graessley and Edwards [Eq. (2)] this concentration dependence determines the scaling exponent $a = 2.22 \pm 0.04$. The exponent lies in the region of the binary contact models and clearly excludes the packing model ($a = 3$).

Having determined the scaling exponent a, we may now examine the scaling with respect to l_K or C_∞, respectively. The temperature dependence of C_∞ for PE has recently been studied by Boothroyd et al. using SANS on PE melts [17]. In a similar T range as in our experiments, they find

$d(\ln C_\infty)/dT = -1.1 \pm 2 \times 10^{-3} \, \mathrm{K}^{-1}$. Finally, using $d(\ln \rho)/dT = -0.7 \times 10^{-3} \, \mathrm{K}^{-1}$ [18], Eq. (2) predicts

$$d^2(T) \approx C_\infty^{-0.44} \rho^{-1.22}$$

$$\approx \exp[+(1.3 \pm 0.1)10^{-3} T] \qquad (4)$$

This has to be contrasted with the experimental result of $d(\ln d^2)/dT = 2.4 \pm 0.4 \times 10^{-3} \, \mathrm{K}^{-1}$ [Eq. (3)], which is nearly three standard deviations larger. In order to visualize the discrepancy, we have included the theoretical prediction of Eq. (4) also in Fig. 3 as a solid line. Experimental results and scaling predictions are clearly apart.

Concerning the topological calculations of Iwata and Edwards [13], a direct quantitative comparison would require extensive numerical calculations using molecular parameters of the two polymers. These calculations have not been made, leaving us with a qualitative comparison. The model predicts $G_N^0 \approx \phi^a$ with $1.97 \lesssim a \leq 2.2$, enveloping our value of $a = 2.22 \pm 0.04$. However, at high contour length density as realized in PEB, the exponent would be expected to be more close to $a = 2$. The topological interaction parameter $\bar{\gamma}$, measuring the capability of a chain to entangle, is predicted to be a monotonically increasing function of C_∞, or the entanglement distance d should monotonically decrease with increasing C_∞. On a qualitative level we observe such a behavior.

References

1. Ferry JD (1980) Viscoelastic Properties of Polymers. Wiley, New York
2. De Gennes PG (1971) J Chem Phys 55:572
3. Doi M, Edwards SF (1978) J Chem Soc Faraday Trans 2 74:1789; (1978) ibid 74:1802; (1978) ibid 75:28
4. Rouse PE (1953) J Chem Phys 21:1273
5. De Gennes PG (1967) Physics (Long Island City, NY) 3:37
6. De Gennes PG (1981) J Phys (Paris) 42:735
7. Ronca G (1983) J Chem Phys 79:1031
8. Graessley WW, Edwards SF (1981) Polymer 22:1389
9. Kavassalis T, Noolandi J (1988) Macromolecules 21:2869
10. Colby RH, Rubinstein M (1990) Macromolecules 23:2753
11. Edwards SF (1967) Proc Phys Soc 92:9
12. De Gennes PG (1974) Phys Lett (les Ulis Fr) 35:L-33
13. Iwata K, Edwards SF (1989) J Chem Phys 90:4567
14. Mezei F (1980) In: Mezei F (ed.) Lecture Notes in Physics, vol 128. Springer, Berlin, Heidelberg, New York
15. Richter D, Fetters LJ, Huang JS, Farago B, Ewen B (1992) Macromolecules 25:6156

16. Richter D, Farago B, Fetters LJ, Huang LS, Ewen B, Lartigue C (1990) Phys Rev Lett 64:1389
17. Boothroyd AT, Rennie AR, Boothroyd CB (1991) Europhys Lett 15:715
18. Christ B, Tanzer JD, Graessley WW (1987) J Polym Science Poly Phys 25:545

Authors' address:

D. Richter
KFA Jülich
IFF Postfach 1913
5170 Jülich, FRG

Progress in Colloid & Polymer Science   Progr Colloid Polym Sci 91:135–137 (1993)

Solvent reorientation dynamics in Aroclor/polymer solutions

A. K. Rizos, K. L. Ngai[1]), and G. Fytas

Foundation for Research and Technology Hellas and University of Crete Department of Chemistry, Heraklion, Crete, Greece
[1]) Naval Research Laboratory, Washington, USA

Abstract: Dynamic light scattering experiments have been conducted in order to probe the orientational dynamics of the solvent Aroclor (A1248) in solutions containing either 1,2- or 1,4-polybutadiene (PB). The measurements of the orientational time correlation functions $C_{VH}(t)$ were carried out over the temperature range 262–233 K and polymer concentrations of up to 25%. $C_{VH}(t)$ is described by a unimodal distribution of solvent relaxation times near T_g and the width of the distribution increases with the addition of the polymer. The solvent reorients faster in the presence of the polymeric solute. These results are described in terms of the Kohlrausch–Williams–Watts (KWW) relaxation function with exponent $\beta = 0.64$ for A1248, 0.58 for a 20% 1,2-PB ($M_w = 4300$)/A1248 solution and discussed in terms of the coupling scheme of relaxation.

Key words: Solvent modification – dynamic light scattering – coupling model

Introduction

Extensive viscoelastic studies of the dynamics of macromolecules in dilute solutions have been reported over the last several years by Schrag and coworkers [1–3]. Every system examined has shown a finite limiting value η'_∞ for the real part of the complex viscosity η^*, contrary to the predictions of simple bead–spring model theories. Furthermore, η'_∞ is sensitive to the specific side group incorporated in the polymer. At present, there is no theory which predicts the observed viscoelastic properties, and the physical origin of η'_∞ is still not clear.

The alteration of the solvent properties by the polymer has been studied by a variety of techniques that probe the solvent dynamics selectively. Nuclear magnetic resonance (NMR), forced Rayleigh scattering [3], oscillatory electric birefringence [2], depolarized Rayleigh spectroscopy (DRS) [4] and photon correlation spectroscopy (PCS) [5] were employed to examine the translational and rotational diffusivity of Arochlor (A1248) in polystyrene (PS), polyisoprene (PI) and polybutadiene (PB) solutions.

In this context, we have attempted to study the effect of the variation of the polymeric structure on the modification of the solvent dynamics by the presence of the polymer by extending our previous measurements (PCS) to two new systems of 1,2-polybutadiene (1,2-PB) and 1,4-polybutadiene (1,4-PB) in A1248. In this paper, we report the main experimental findings of the A1248 spectra in solutions of polymers, point out the unexpected results of the 1,2-PB/A1248 system and explain the observed properties in the framework of the coupling model of relaxation.

Experimental results

The time correlation function $G(t)$ of the depolarized scattering intensities for the neat solvent, the PS ($M_w = 9.5 \cdot 10^4$)/A1248, 1,4-PB ($M_w = 2500$)/A1248 and 1,2-PB ($M_w = 4300$)/A1248 solutions was measured at different temperatures near and above T_g in the time range 10^{-6}–10^2 s. The net experimental correlation function $C_{VH}(t) = [G(t)/A - 1]^{1/2}$, where A is the baseline, has a nonexponential shape that is well represented by

Table 1

System	β	T_g (°C)
1,4-PB	0.4	− 92
1,2-PB	0.33	− 26
PS	0.37	100
A1248	0.64	− 44
1,4-PB/A1248	0.50	− 58
1,2-PB/A1248	0.58	
PS/A1248	0.49	− 41

a Kohlrausch–Williams–Watts (KWW) function:

$$C_{VH}(t) = \alpha \exp[-(t/\tau^*)^\beta] . \qquad (1)$$

The factor α, which is the fraction of the depolarized scattering intensity arising from fluctuations in the optical anisotropy, with correlation times longer than about 10^{-6} s, was found to be equal to 1 for all the polymer/A1248 systems studied. Hence, the fluctuations of the optical anisotropy near T_g relax within the time window of the PCS. The distribution parameter β and relaxation time τ are defined from the fit of Eq. (1) to the experimental $C_{VH}(t)$. The β values for the different polymer/A1248 systems studied are listed in Table 1.

The reorientation time increases in PS/A1248 solutions and decreases in 1,4-PB/A1248 solutions, as it was observed before. By contrast, for the 1,2-PB/11248 system the relaxation times become faster, despite the fact that the T_g of 1,2-PB lies about 19 K above the T_g of A1248. This intriguing feature simply cannot be explained by the conventional free-volume model.

Discussion

Recently, an alternative theory has been proposed for neat solvent relaxation that is based on the Adams–Gibbs theory [6], but modified according to the coupling model [7, 8], which was extended to incorporate some essential new elements. The system studied can be modelled [9] by independent, equivalent, but distinguishable, cooperative rearranging regions (CRRs). The coupling scheme introduces cooperativity between the relaxing CRRs, and the autocorrelation function $q(t)$ of the transition of a CRR from one configuration to

another will have the form of Eq. (1), where the coupling parameter $n = 1 - \beta$ captures the strength of the coupling between the CRRs and it is a measure of the slowing down of the relaxation rate by dynamic constraints or correlations between relaxing CRRs. After a time scale $t_c \equiv \omega_c^{-1}$, characteristic of the mutual interactions, the relaxation rate is slowed down to have the self-similar time dependence of $W(t) = W_0(\omega_c t)^{-n}$. The experimentally determined relaxation time τ^* is related to the primitive relaxation time τ_0 according to

$$\tau^* = [(1 - n)\omega_c^n \tau_0]^{1/(1-n)} . \qquad (2)$$

The dissolved polymer modifies the A1248 reorientation in at least two different ways. Firstly, due to composition fluctuations there are heterogeneities in the environments and, hence, a broadening of the relaxation spectrum. Secondly, the presence of the polymer can modify the coupling between the CRRs and effectively change the coupling parameter n. A direct application of Eq. (2) to the experimental τ^* values for the undiluted A1248 and polymers yields the τ_0 values of Fig. 1. The faster reorientational dynamics of the 1,2-PB/A1248 solution is a direct consequence of the fact that the primitive relaxation τ_0 of 1,2-PB is faster as compared to the corresponding value measured for A1248 (see Fig. 1). Therefore, the presence of 1,2-PB induces the observed faster reorientation of A1248. The PS and 1,4-PB primitive relaxation τ_0 values are, respectively, slower and

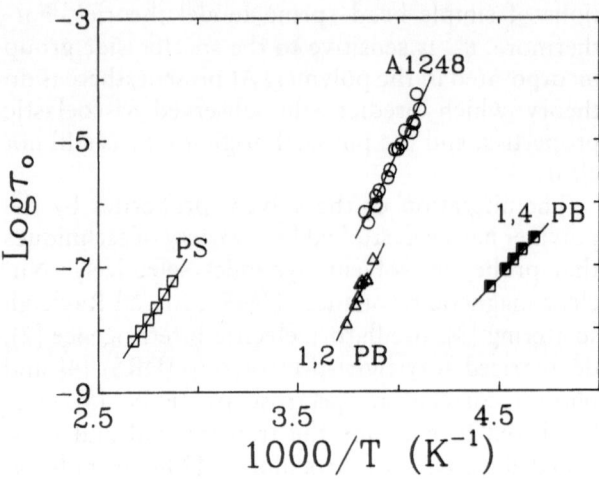

Fig. 1. Temperature dependence of relaxation times of log τ_0/s for PS, A1248 in the neat state, 1,4-PB and 1,2-PB

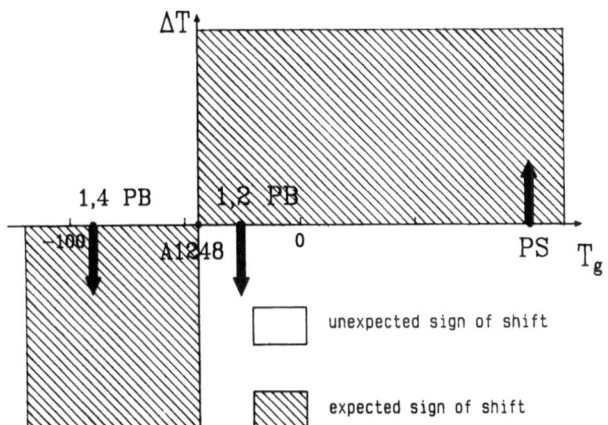

Fig. 2. Schematic representation of the observed trends with variations in the polymer structure

faster than the corresponding ones for A1248, also in agreement with observed trends in the polymer/A1248 solutions. This solvent modification expressed in temperature shift ΔT necessary to superimpose the τ^* values (see Fig. 4 of ref. [5]) of A1248 in the presence of polymer to the corresponding τ^* values of neat A1248 is shown schematically in Fig. 2. The slower reorientation of A1248 in PS/A1248 solutions and the corresponding faster one in 1,4-PB/A1248 solutions have been verified by NMR in a very recent study [10].

Conclusion

In this paper we have provided new evidence of polymer-induced modification of solvent dynamics. In this context we have varied the polymeric structure by performing dynamic light scattering measurements of the 1,2-PB and PI/A1248 systems. Although the T_g of 1,2-PB is higher than that of A1248, yet we observed that addition of 1,2-PB to A1248 makes relaxation of A1248 in the polymer solution faster. This anomalous behavior can be explained by the coupling model in providing a consistent determination of the relevant relaxation parameters.

References

1. Lodge TP, Schrag JL (1984) Macromolecules 17:352
2. Morris RL, Amelar S, Lodge TP (1988) J Chem Phys 89:6523
3. Meerwall von D, Amelar S, Smeltzy MA, Lodge TP (1989) Macromolecules 22:295
4. Fytas G, Rizos A, Floudas G, Lodge TP (1990) J Chem Phys 93:5096
5. Rizos A, Fytas G, Lodge TP, Ngai KL (1991) J Chem Phys 95:2980
6. Adams G, Gibbs JH (1965) J Chem Phys 43:139
7. Ngai KL, Rendel RW, Rajagopal AK, Teitler S (1986) Ann NY Acad Sci 484:150
8. Rajagopal AK, Ngai KL, Teitler S (1988) J Chem Phys 88:5086
9. Ngai KL, Rendel RW, Plazek DJ (1991) J Chem Phys 94:3018
10. Gisser DJ, Ediger MD (1992) Macromolecules 25:1284

Received January 20, 1992;
accepted June 2, 1992

Authors' address:

A. K. Rizos
Foundation for Research
and Technology Hellas
P.O. Box 1527
71 100 Heraklion, Crete, Greece

Dynamic light scattering and viscoelasticity in polymer solutions

C. H. Wang

Department of Chemistry, University of Nebraska, Lincoln, USA

Abstract: The dynamic light-scattering spectrum from a polymer solution is shown to be related to the cooperative diffusion and longitudinal stress relaxation modes. Experimental results are presented to corroborate the theoretical analysis.

Key words: Dynamic light scattering – polymer solution – cooperative diffusion – viscoelasticity

Introduction

Dynamic light scattering from a dilute polymer solution with a monodisperse molecular weight polymer is essentially characterized by a single exponentially decaying time autocorrelation function of the intensity of the scattered light. If the amplitude of the scattering vector q is sufficiently small such that $qR_g < 1$, where R_g is the mean radius of gyration of macromolecule in solution, then the decay rate constant is proportional to the translational diffusion coefficient of the polymer in the solvent.

Increasing the polymer concentration slightly causes the diffusion coefficient to either decrease or increase with increasing polymer concentration, depending on the solvent quality. For good solvents, the diffusion coefficient increases with increasing concentration, but it decreases for poor solvents. In the dilute concentration regime, there is no significant change in the shape of the autocorrelation function when the polymer concentration is changed. However, increase in the polymer concentration beyond the overlap concentration, defined by $C^* = M/(\frac{4}{3}\pi R_g^3)N_A$, results in major deviations from the single exponential decay in the time correlation function [1–5]. Here N_A is the Avogadro's number, M is the molecular weight. The overlap concentration C^* separates the dilute from the semidilute regime. Above C^*, neighboring polymer chains begin to overlap, and the hydrodynamic interaction is screened. The effect results in a complex concentration dependence for

the diffusion coefficient. Using a transient gel model, de Gennes developed a scaling theory for the semidilute solution in thermodynamically good solvents [6, 7]. The central prediction of the theory is that a single characteristic length exists in such a system and, consequently, only one single dynamic process, characterized by a cooperative diffusion coefficient (D_c) is expected.

Experimentally, it is found, however, that although the cooperative diffusion process dominates in the semidilute solution, other modes are also present. These modes lead to deviation from a single exponential decay of the time correlation function, becoming more pronounced as the solvent quality is decreased [8]. Brochard modified de Gennes' theory by allowing the transient gel to relax, and obtained for the time correlation function a q^2-dependent cooperative diffusion mode and a single q-independent relaxation viscoelastic mode in their calculation [9]. The q^2-dependent cooperative diffusion mode was referred to as the fast mode and the q-independent relaxation viscoelastic mode as the slow mode, associated with the chain disengagement relaxation.

Brown, Stepanek and coworkers have analyzed the shape of the time correlation function of the semidilute polymer solution [10, 11] using the CONTIN [12] program to find the distribution of relaxation times. They have found a bimodal distribution in the relaxation time spectrum, with one mode having the relaxation rate proportional to q^2 and the other consisting of a group of closely spaced q-independent modes having relaxation times stretching from the fast to the slowest viscoelastic mode. The diffusion coefficient associated with the q^2-dependent mode in the semidilute solution is associated with the cooperative diffusion coefficient, D_c. They further examined similarities between the relaxation time distributions for polystyrene in semidilute solutions in the θ solvent, dioctyl phthalate (DOP), obtained from dynamic light scattering and dynamic mechanical (shear) measurements [3, 11]. They have found that the dynamic processes probed at long times from both techniques are characterized by a similar dynamic range of relaxation; however, as the solvent quality changes, the distribution of relaxation times from dynamic light scattering appears to be more sensitive in the slow part of the spectrum [11]. The empirical approach ventured by Brown, Stepanek and coworkers has motivated this author to undertake a theoretical analysis to examine the effect of viscoelasticity on the DLS spectrum [13].

In this paper, we review the theoretical results obtained recently; and to support the theoretical result, we provide preliminary experimental results of the quasielastic light-scattering study of the semidilute polymer solution of polystyrene (PS) in diethyl phthalate (DEP) obtained in our laboratory. At room temperature, the second virial coefficient of the solution is positive; DEP is, thus, a good solvent. A comprehensive account of the experimental study of the solution of PS in DEP over a wide concentration range with variable PS molecular weights will be the subject of another publication.

Experimental

Polystyrene samples were purchased from Toya Soda. Solutions were prepared by dissolving polystyrene into diethyl phthalate (Pfaltz and Bauer, Inc.). Dusts in solutions with less than 1 wt% were removed by 0.25 μ Millipore-filters and for solutions with more than 1% by centrifuge. Dynamic light scattering was carried out with an experimental setup described previously [14]. The light source was a 632.8 nm He–Ne laser with a power about 40 mM. A correlator with multiple-sample times was used to obtain the time correlation function in the logarithmic time scale. The Zimme plot was also obtained to evaluate the second virial coefficient.

Results and discussion

For a polymer solution in the dilute and semidilute regions, the refractive index n is modulated more strongly by the concentration fluctuation than by the density fluctuation. As a result, the dynamics of concentration fluctuations dominate the light-scattering spectrum. Using generalized hydrodynamic equations, the light-scattering spectrum due to concentration fluctuations is recently shown to be [13]:

$$I(q, \omega) = \left(\frac{\partial \varepsilon}{\partial \rho_2} \right)^2 < |\delta \rho_2(q)|^2 > S(q, \omega) . \quad (1)$$

The quantity $S(q, \omega)$ is the dynamic structural factor given by

$$S(q, \omega) = \mathrm{Re} \left(\frac{\hat{m}(\omega) + \xi}{\Delta(\omega)} \right) \quad (2)$$

where ω is the angular frequency, and ξ is the fraction coefficient on the polymer chain due to solvent molecules, $\Delta(\omega)$ is a dispersion function defined in ref. [13] and $\hat{m}(\omega)$ is given by

$$\hat{m}(\omega) = \frac{q^2}{\rho_2^0} \beta \int_0^\infty dt\, e^{-i\omega t} M(t) . \quad (3)$$

Here $\beta = (\rho_2^0/\rho_0)(\partial \rho/\partial \rho_2)_{T,p}$, ρ_2^0 is the concentration of the polymer and ρ_0 is the density of the solution. The quantity β determines the effectiveness of the coupling of the concentration fluctuations to the viscoelasticity of the polymer solution. The coupling is important when there exists a large difference in the partial specific volumes of the polymer and the solvent component in the solution as in this case adding polymer to the solution causes a significant change in local concentration due to the volume change; in contrast, only negligible coupling to the concentration fluctuations is expected if the partial specific volumes of the two components are not much different. In a good solvent, polymer chains are extended, and the solvent molecules strongly solvate the polymer chains. Adding the polymer to the solution with good solvent is not expected to change the solution density appreciably, hence, resulting in a small change of β. On the other hand, we expect a large change in β with concentration for polymer in the θ-solvent. Basically, the β parameter is a measure of the fraction of density fluctuation that makes contribution to the concentration fluctuation dynamic through the viscoelasticity of the polymer solution.

The normalized time correlation function due to concentration fluctuations, $g(q, t)$ is the Fourier transform of the dynamic structural factor. In polymer solutions, the stress modulus $M(t)$ is determined by a distribution of relaxation times. For systems for which the condition, $\omega \tau_i > 1$ is satisfied for all relaxation times τ_i, a recent calculation shows that $g(q, t)$ is given by [15]

$$g(q, t) \equiv \frac{\langle \delta \rho_2(q, t) \delta \rho_2^*(q) \rangle}{\langle |\delta \rho_2(q)|^2 \rangle}$$

$$= A e^{-D_c q^2 t} + \sum_i B_i e^{-t/\tau_i}, \qquad (4)$$

where for q such that $D_c q^2 \gg \tau_i^{-1}$, A and B are given by

$$A = 1 - \frac{\beta M_0}{\xi \rho_2^0 D_c} = 1 - \beta M_0/(\beta M_0 + M_\pi) \quad (5a)$$

$$B_i = \beta M_i/(\beta M_0 + M_\pi). \qquad (5b)$$

Here D_c is the cooperative diffusion coefficient given by $D_c = (M_\pi + \beta M_0)/\rho_2^0 \xi$; M_π is the osmotic modulus; and $M_0 = \sum_i M_i$, M_i being the amplitude of the stress modulus of the ith viscoelastic mode.

Equation (4) shows that the coupling of the concentration fluctuation to the viscoelastic effect results in a bimodel time autocorrelation function of concentration consisting of one single exponential decay associated with cooperative diffusion and a group of relaxation modes consisting of time-independent amplitudes, characterizing the structural (longitudinal stress) relaxation of the binary solution.

To test the theoretical prediction, we have measured the intensity autocorrelation function of the scattered light of polystyrene (PS) in diethyl phthalate (DEP). The function $g(q, t)$ deduced from the experiment is shown in Fig. 1. One notes that $g(q, t)$ takes more than six decades to relax. By subjecting $g(q, t)$ to either the inverse Laplace transform (CONTIN) analysis or to a double stretched exponential function fit, we obtain a bimodal structure: with a nearly single exponential model and a group of modes with a wide distribution of relaxation times. The relaxation rate constant of the single exponential mode is proportional to q^2; however, except for at small q, the mean relaxation time of the broad nonexponential

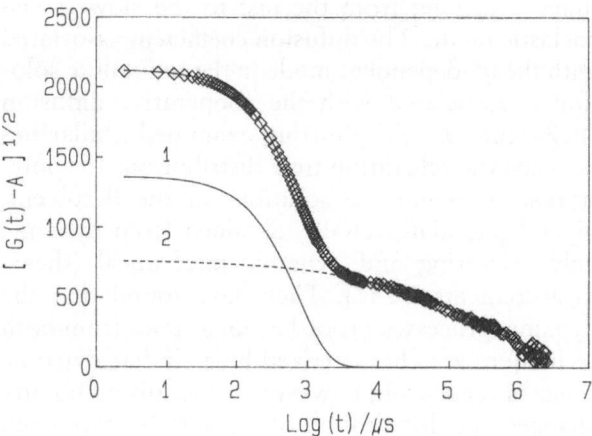

Fig. 1. The measured time correlation function $(G(t))$ of PS in DEP M_w of PS is $2.88 \cdot 10^6$ ($M_w/M_n = 1.03$) and the concentration is 0.1025 gg^{-1}. The top curve is the experimental result, which is resolved into curve 1 (single exponential) and curve 2 (a broad distribution)

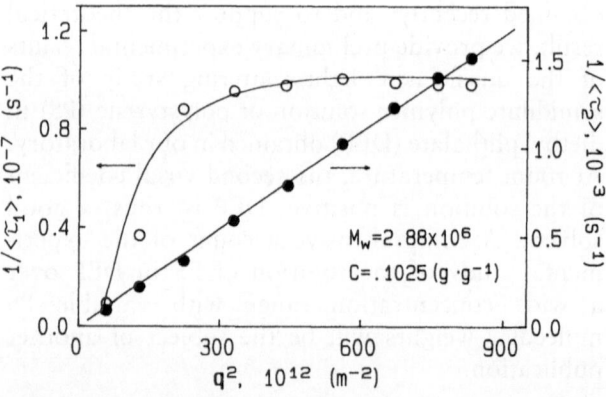

Fig. 2. The q-dependence for the relaxation rate of the single exponential mode (curve 1 in Fig. 1) and for the average relaxation rate of the viscoelastic modes (curve 2 in Fig. 1). The strong q-dependence for the viscoelastic modes at small q is due to q-dependence amplitudes which occur when $D_c q^2 < \tau_i^{-1}$ (ref. [14])

mode is q-independent (Fig. 2). The amplitude of the q^2-dependence made is relate to the osmotic modulus and the mechanical modulus according to Eq. (5a).

These results are qualitatively consistent with the theoretical prediction, that is, the q^2-dependent mode is associated with cooperative diffusion and the broadly distributed q-independent modes are associated with viscoelastic relaxation. To quantitatively test the theory, it is necessary to determine the coupling parameter β and also evaluate the amplitude factors A and B_i, which can be obtained by the intensity and viscoelastic measurements. The β value for the PS/DEP system at 20 °C has been obtained from the density measurement to be $\beta = 5.545 \cdot 10^{-2} \rho_2$. Since viscoelastic modes enter the correlation function through the βM_0 term in the q^2-dependent diffusive mode, the conclusion reached previously about $M_\pi/M_0 \gg 1$ for good solvent solution and $M_\pi/M_0 \simeq 1$ for θ solution is probably incorrect [9], considering the fact that M_π is usually several order of magnitude smaller than the mechanical longitudinal modulus. We expect $M_\pi/M_0 \ll 1$ for both types of solutions. In fact, it is the result of combining M_0 with β that renders it to have a comparable value with M_π. Additionally, work is in progress and the result will be reported elsewhere in greater detail [16].

Acknowledgements

We thank X. Q. Zhang for performing the light scattering experiments the Office of Naval Research for partial financial support of this research.

References

1. Nose T, Chu B (1979) Macromolecules 12:590, 599; Chu B, Nose T (1980) Macromolecules 13:122
2. Adam M, Delsanti M (1985) Macromolecules 18:1760
3. Nicolai T, Brown W, Hvidt S, Heller K (1990) Macromolecules 23:5088
4. Selser JJ (1983) Chem Phys 79:1044
5. Amis EJ, Han CC, Matsushita Y (1984) Polymer 25:650
6. de Gennes PG (1976) Macromolecules 9:587
7. Brochard F, de Gennes PG (1977) Macromolecules 10:1157
8. Brown W, Stepanek P (1988) Macromolecules 21:1791
9. Brochard F (1983) J Phys 44:39
10. Brown W, Johnsen RM, Stepanek P, Jakes J (1988) Macromolecules 21:2859
11. Brown W, Nicolai T, Hvidt S, Stepanek P (1990) Macromolecules 23:357
12. Provencher SW (1979) Macromol Chem 180:201
13. Wang CH (1991) J Chem Phys 95:3788
14. Xia JL, Wang CH (1991) J Chem Phys 94:3229
15. Wang CH (1992) Macromolecules 25:1524
16. Wang CH, Zhang XQ Macromolecules (to be published)

Received January 22, 1992;
accepted June 18, 1992

Author's address:

C. H. Wang
Department of Chemistry
University of Nebraska
Lincoln, NE 68588-0304, USA

Progress in Colloid & Polymer Science   Progr Colloid Polym Sci 91:142–145 (1993)

Dynamics of polymer chains from expanded coils to the collapsed state

B. Chu, J. Yu, and Z. Wang

Department of Chemistry, State University of New York at Stony Brook, Long Island, USA

Abstract: Static and dynamic laser light scattering and viscometry have been used to study the dynamics and structure of coil-to-globule transition. With the prism-cell laser light scattering spectrometer, very small scattering angles of the order of 3° could be reached which enabled us to determine the pure translational motion of large polymer chains. The contraction size of a polymer chain during the collapse process has been measured in terms of the radius of gyration, the hydrodynamic radius and the intrinsic viscosity. In the thermodynamically stable region, the coil size obeyed the scaling theory. However, a kinetic study of the collapse has demonstrated a two-stage collapse process in agreement with the theoretical prediction of Grosberg et al. Experimental results of the coil-to-globule transition from polystyrene in cyclohexane and from a polyelectrolyte (sodium polystyrene sulfonate) in aqueous solution are presented.

Key words: Laser light – scattering – polymer chain collapse – coil-to-globule transition – polyelectrolyte – polystyrene

Introduction

The contraction of a polystyrene (PS) chain in cyclohexane (CY) was observed with decreasing temperature below the θ temperature and the size decrease obeyed the scaling theory [1]. However, upon reaching a collapsed state, we should expect the coil density to be close to that of the bulk polymer. Unfortunately, the degree of contraction reported in [1] reached only of the order of 10%–15%, instead of an expected factor of say 5. This small decrease in size certainly does not signal the usual definition of a collapsed state [2]. We now present a continuation of our studies on the coil-to-globule transition both in terms of the coil dynamics and of the chain size contraction.

Experimental

A coil collapsed state can be observed only at very dilute solution concentrations. Laser light scattering (LS) and viscometry are appropriate techniques for studying the collapse in terms of the radius of gyration (R_g), the hydrodynamic radius (R_h) and the intrinsic viscosity ($[\eta]$). Dynamic LS measurements have to be performed in the range of $KR_g < 1$ with K being the magnitude of the scattering vector in order to obtain the pure translational motion of a polymer chain without interference of chain internal motions. A unique prism-cell light scattering spectrometer with a light-beating efficiency (β) value of about 0.9 [1] was used. The spectrometer could also perform self-beating experiments down to a scattering angle of $\sim 3°$. A capillary viscometer with a flow time measurement precision of 0.001 s [3] was used for $[\eta]$ measurements.

Results and discussions

Thermodynamically stable region

The expansion factor α of a polymer chain exhibits a universal behavior by using the scaled variable $\tau M^{1/2}$ with $\tau(= 1 - T/\theta)$ being the reduced temperature. Figure 1 shows variations of the scaled expansion factor $\alpha^3 \tau M_w^{1/2}$ as a function

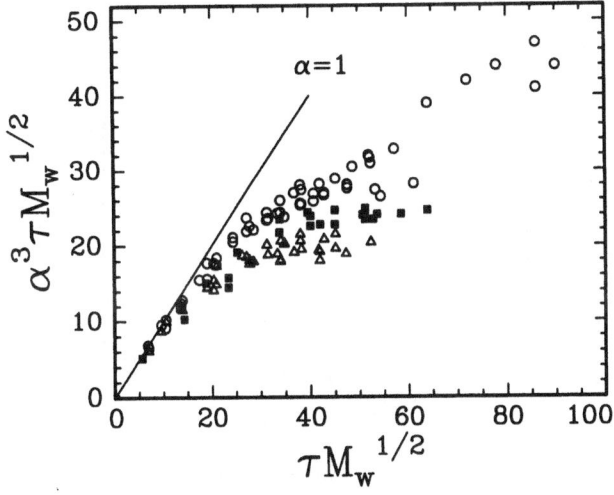

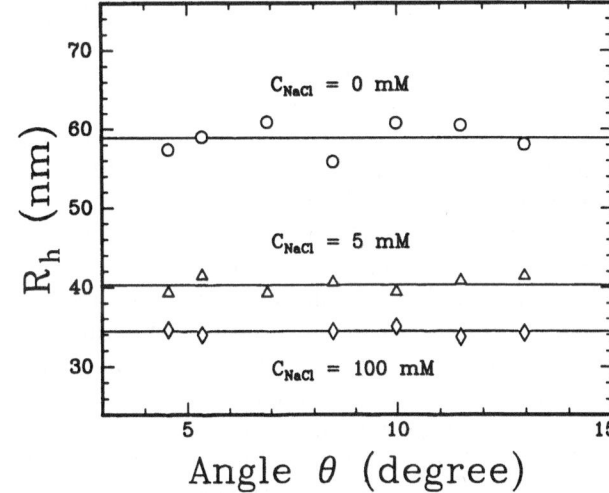

Fig. 1. Universal curves of scaled expansion factor $\alpha^3 \tau M_w^{1/2}$ vs. reduced temperature $\tau M_w^{1/2}$. Circles, triangles and filled squares denote the scaled expansion factors calculated from the hydrodynamic radius R_h, the radius of gyration R_g and the intrinsic viscosity $[\eta]$, respectively [1, 2]

Fig. 2. Plots of hydrodynamic radius of polyelectrolyte NaPSS vs. scattering angle at different added salt concentration with NaPSS concentration $\sim 5 \cdot 10^{-5}$ g/mL. Measurements were performed at 15 °C

of scaled reduced temperature $\tau M_w^{1/2}$ in the thermodynamically stable state. A plateau region in $\alpha^3 \tau M_w^{1/2}$ was observed, suggesting a "collapsed" regime according to the scaling theory [4]. The values of $\alpha^3 \tau M_w^{1/2}$ at the plateau regime for $\alpha_s (= R_g(T)/R_g(\theta))$, $\alpha_h (= R_h(T)/R_h(\theta))$ and $\alpha_\eta (= ([\eta](T)/[\eta](\theta))^{1/3})$ were about 20, 43 and 24 $g^{1/2}/mol^{1/2}$, respectively. Experimental ratios of $\alpha_h^3 : \alpha_\eta^3 : \alpha_s^3 = 2.2 : 1.2 : 1$ were obtained. Thus, we have $\alpha_\eta^3 \approx \alpha_s^2 \cdot \alpha_h$. The proposed relations for the different expansion factors were $\alpha_\eta^3 = 1.7 \alpha_s^2 \cdot \alpha_h$ and $\alpha_\eta^3 = \alpha_s^2 \cdot \alpha_h$ for a polymer chain in the globular state and in the θ state, respectively [5]. Thus, the plateau values suggest that we have not yet reached the totally collapsed state.

We noticed that the globule size at the plateau region to the ideal size at θ temperature observed at thermal equilibrium was only about 0.7 for the PS/CY system under the best of conditions. The 0.7 value is far from that of a totally collapsed state of a densely packed coil. Although we have reached a contracted state according to the scaling theory, we have certainly not reached a small enough contraction to convince ourselves that we have truly achieved the collapsed state. Some biological macromolecules, e.g., proteins, show their abilities of folding the chain into a dense globule, indicating the accessibility of a highly compact state. Some preliminary measurements on the condensation of

a polyelectrolyte in solution is also presented. The sodium polystyrene sulfonate (NaPSS) has an extended chain conformation in the aqueous solution without added salt due to the electrostatic repulsion between the same charges in the polyelectrolyte chain. With the addition of salt, the counterion condensation causes the NaPSS chain to collapse. Figure 2 shows the contraction in R_h of a NaPSS sample (with $M_w = 1.2 \cdot 10^6$ g/mole and $M_w/M_n \leq 1.10$) by adding NaCl to the solution. A contraction ratio of about 0.58 (vs. 0.7 for the PS/CY system) was achieved by adding NaCl to a dilute NaPSS aqueous solution with a final NaCl concentration of 100 mM.

Kinetic study

Unlike the scaling theory, theoretical work on the kinetics of the collapse process suggested two successive stages, i.e., sausages and compact spheres [2, 6, 7]. Here, we present our experimental observations on the kinetic behavior of a polymer chain collapse.

A highly dilute ($9.6 \cdot 10^{-6}$ g/mL) solution of PS ($M_w = 8.60 \cdot 10^6$ g/mole) in CY was used [8]. To reach the designated temperature, an abrupt temperature change was used to "quench" the solution from the θ temperature to the final temperature. It was different from the temperature change used in the previous collapse studies in which the temperature was decreased from the θ temperature step by step until aggregation occurred. In the present experimental setup, the time for the polymer solution to reach the new designated temperature quenched from the θ temperature was about 15–20 min without perturbing the solution. An interesting two-stage collapse process has been observed at temperatures slightly lower than those temperatures at which the polymer solution was thermodynamically stable. At these (lower) temperatures, after about 20 min for thermal equilibration of the entire polymer solution to the new designated temperature, the polymer chain reached a contracted state whose behavior could be fitted to the master curve of the scaling theory as shown in Fig. 1. This contracted state lasted for a certain time which was temperature- and concentration-dependent. After that time, aggregation of polymer chains started. In addition to the observation of bigger particles which was due to the aggregation, a smaller particle size was also observed. While the size of the bigger particles continued to increase, the size of the smaller particles decreased with increasing time and then approached a relatively stable value for another certain time. Finally, the aggregates dominated the scattering intensity. Figure 3 shows the time dependence of the hydrodynamic size and the amplitude of the smaller particles. At the second stage, the globule size contracted to about 0.38 of the ideal coil size at the θ state. The absolute excess scattered intensity could be used to exclude the strong molecular weight frationation effect due to phase separation.

The present kinetic observation of a two-stage process suggests that Fig. 1 shows only the first stage of the collapse process. The observed two-stage process resembles the collapse process proposed by the kinetic theory [6, 7], where both stages could remain in a stable one-phase regime. The difference between experiment and theory is that finite polymer concentrations are required in order for the phenomenon to be observable.

In conclusion, the present study demonstrated experimental observations on both static and kinetic properties of the polymer collapse. In the thermodynamically stable regime, i.e., in the equilibrium one-phase region, the behavior of the PS/CY system could be described by the scaling theory; but the coil contraction which can be achieved by the PS/CY system is limited under the

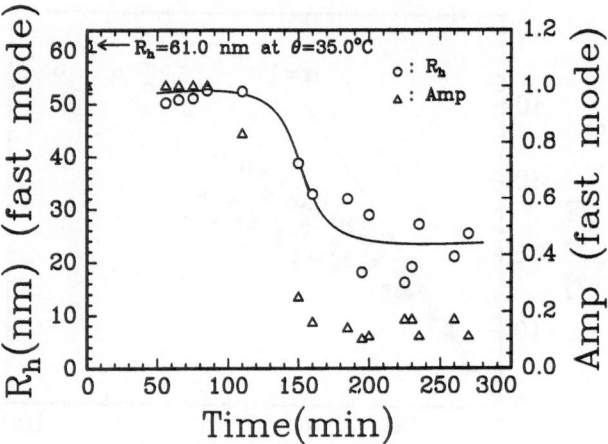

Fig. 3. Time dependence of hydrodynamic size and amplitude of PS in CY during the two-stage collapse process. PS concentration = $9.6 \cdot 10^{-6}$ g/mL and temperature = 29.0 °C. Over a time period of one hour after the solution reached thermal equilibrium, the polymer chain achieved the first stage of collapse. Over a time period of 200–270 min after thermal equilibrium, the polymer chain reached a second stage of the collapse process

accessible experimental conditions. Kinetically, a two-stage collapse process was observed which was similar to the theoretical prediction of [6, 7] and could be observed only in the presence of phase separation [8]. Further detailed experiments are under way.

Acknowledgements

We gratefully acknowledge the support of this work by the Polymers Program, National Science Foundation (DMR 8291968).

References

1. Park I, Wang Q, Chu B (1987) Macromolecules 20:1965–1975, 2883–2840; Chu B, Xu R, Zuo J (1988) Macromolecules 21:273–274; and references therein
2. Grosberg AYu, Kuznetsov DV (1992) Macromolecules 25:1996–2003, and earlier three references in volume 25
3. Chu B, Wang Z (1989) Macromolecules 22:380–383
4. Park IH, Ph.D. Thesis (1986) State University of New York at Stony Brook, Stony Brook, New York, USA. The thesis lists the detailed experimental data on the coil contraction
5. Perzynski P, Adam M, Delsanti M (1982) J Physique 43:129–135
6. Grosberg A, Nechaev S, Shakhnovick E (1988) J Phys France 49:2095–2100; and references therein

7. de Gennes P (1985) J Physique Lett 46:L639–L642
8. Yu J-Q, Wang Z-L, Chu B (1992) Macromolecules 25:1618–1620

Received January 26, 1992; accepted June 24, 1992

Authors' address:

Prof. Benjamin Chu
Department of Chemistry
State University of New York at Stony Brook
Long Island, NY 11794-3400, USA

Progress in Colloid & Polymer Science

Progr Colloid Polym Sci 91:146–148 (1993)

Chain fragmentation and fragment diffusion at the glass transition

G. P. Hellmann[1]), E. H. Hellmann[2]) and A. R. Rennie[3])

[1]) Deutsches Kunststoff-Institut, Darmstadt, FRG
[2]) Max-Planck-Institut für Polymerforschung Mainz, FRG
[3]) Institut Laue-Langevin Grenoble, France

Abstract: Polymer chain diffusion was measured by SANS using a new technique. The chains of a copolycarbonate were fragmented, in blends with tetramethyl-polycarbonate, in the glassy state, so that the split chains remained first conformationally intact. The subsequent diffusion of the chain fragments alters the structure factor of the blend. This was monitored by SANS and interpreted in terms of center-of-mass diffusion.

Key words: Chain fragmentation – chain diffusion – polymer blends – neutron scattering

Chain diffusion is slow near the glass transition, so that methods to measure it must respond to short diffusion distances. The wave vector range of small-angle neutron scattering (SANS) is perfectly suited. On the one hand, the spatial resolution is better than for light scattering and, on the other hand, the distances covered are still long enough to ensure that diffusive motions are measured, and not local motions inside chemical groups. The design of diffusion experiments that make full use of SANS, however, is not easy. The experiments must be of the approach-to-equilibrium type, meaning that a well-defined non-equilibrium state is needed to start from.

One possibility is to start with a blend A/B that has a structure factor S_f (f is the volume fraction of A) given by (for ideally polydisperse polymers)

$$\frac{f(1-f)}{S_f} = 1/\bar{V} - 2\chi f(1-f) + \bar{C}^2 q^2/12 \,,$$

$$1/\bar{V} = (1-f)/V_A + f/V_B \,,$$

$$\bar{C}^2 = (1-f)C_A^2 + fC_B^2 \,, \tag{1}$$

and to induce a non-equilibrium state by suddenly changing conditions, so that one of the parameters is no longer in equilibrium. The chain stiffness parameter $\bar{C}^2$ cannot be changed, so that the choices are changes of the interaction parameter χ or of the chain volume $\bar{V}$. The relaxation $\chi_0 \to \chi_\infty$ or $\bar{V}_0 \to \bar{V}_\infty$ from the non-equilibrium state (0) into the new equilibrium state (∞) can then be monitored by SANS.

Changing χ, by a jump in temperature or pressure, is the conventional alternative [1]. In this study, $\bar{V}$ was changed, by controlled fragmentation of one of the polymers. The technique is illustrated in Fig. 1.

– If chains are split in the liquid state, the fragments diffuse apart, right after each dissociation event. Correlations between fragments are lost immediately (upper line in Fig. 1). The time constant τ of the fragmentation reaction can be measured by SANS.

– If chains are split in the glassy state, however, the diffusion of the fragments is slow. There will be less or no loss in correlation between fragments (lower line in Fig. 1). Starting from this state, the conversion $0^* \to \infty$ via chain fragment diffusion (CFD) can be measured by SANS and, thereby, the time constant δ.

Experiments were made with copolymers $PC(T_x A_{1-x})$ that are related to polycarbonate

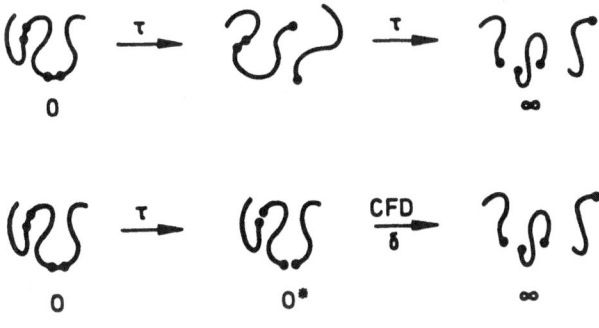

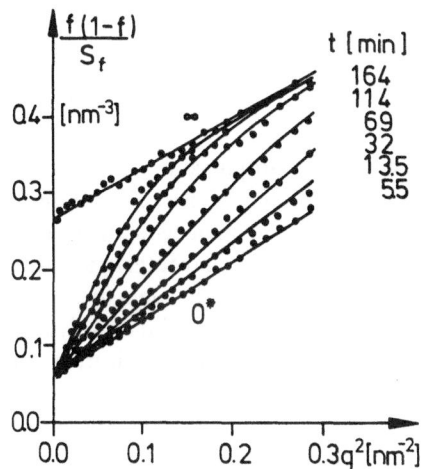

Fig. 1. Chain fragmentation and chain fragment diffusion (CFD)

(comonomer CA) and contain thermosensitive sites (comonomer CT). The copolymers were homogeneously blended with polycarbonate (PC) or tetramethylpolycarbonate (TMPC) [2].

The comonomers CT split up, in a thermally activated reaction of an Arrhenius temperature dependence, with a time constant $\tau = 1\,\mathrm{h}$ at $T^* = 172\,°\mathrm{C}$. The state 0* was prepared at $T < T_g$ in blends with TMPC, of a low copolymer content f (where $T^* < T_g$).

Figure 2a shows a set of Zimm curves caused by the CFD process $0^* \to \infty$. The copolymer $PC(T_{0.16}A_{0.84})$ had a weight average degree of polymerization of $M_w^0 = 31.200$ (corresponding to the degree of polymerization $\lambda_w^0 = 120$) before and $M_w^\infty = 2.850$ ($\lambda_w^\infty = 11$) after fragmentation. In both the states, the polydispersity was nearly ideal ($M_w/M_n \cong 2$). The copolymer was blended, at a concentration $f = 0.12$, with the matrix polymer TMPC ($M_w = 23.600$, $\lambda_w = 76$). The state 0* was established by total fragmentation of the copolymer at $150\,°\mathrm{C} = (T_g - 36\,\mathrm{K})$. The final state ∞ was established by heating the blend to $200\,°\mathrm{C}$ for 10 min.

During the chain fragmentation process seen in Fig. 2a, the structure factor $S(t)$ of the copolymer changes as [2]

$$S(t) = S_\infty + (S_0 - S_\infty)$$
$$\times \langle \exp - D_w (\lambda_w^\infty/\lambda)\, q^2 t \rangle_w^2 , \qquad (2)$$

where D_w is the diffusion coefficient of the weight average fragments with λ_w^∞. Equation (2) was derived under the assumptions (i) that the fragments

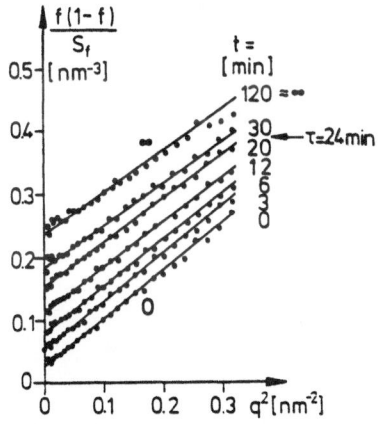

Fig. 2. Zimm diagrams of (a) fragment diffusion in a blend $PC(T_{0.16}A_{0.84})/TMPC$ at $177\,°\mathrm{C} = (T_g - 9\,\mathrm{K})$, (b) fragmentation in a blend $PC(T_{0.16}A_{0.84})/PC$ at $180\,°\mathrm{C} = (T_g + 36\,\mathrm{K})$

were, in their internal conformations, Gaussian subchains, and (ii) that the fragments diffuse apart only by center-of-mass diffusion. Of the curves in Fig. 2a, that for the initial state 0*, where the copolymer chains are split but conformationally still intact, and that for the final state ∞, where the fragments are uncorrelated, are described by Eq. (1).

In the intermediate curves of Fig. 2a, the conversion $0^* \to \infty$ lags behind at lower q values. This curve shape is characteristic of diffusion processes. Diffusion coefficients in the very low range 10^{-22}–10^{-18} cm^2/s were measured, in the temperature range $(T_g - 24\,\mathrm{K}) \to (T_g + 14\,\mathrm{K})$. (For data evaluation see ref. [2].) The correlation time δ in Fig. 1 is, approximately, given by

$$1/\delta = D_{\mathrm{w}} q^2 \ . \tag{3}$$

Figure 2b shows, for comparison, what happened at $T > T_{\mathrm{g}}$, in blends with PC (where $T^* > T_{\mathrm{g}}$). The fragmentation proceeds in a relatively fluid state, so that the characteristic time τ for the fragmentation reaction is measured, instead of δ. Since the chain volume of the copolymer drops monotonously, during fragmentation, a set of parallels is obtained (Eq. (1)).

References

1. Strobl GR (1985) Macromolecules 18:558
2. Hellmann EH, Hellmann GP, Rennie AR (1991) Macromolecules 24:3821

Received January 8, 1992;
accepted June 12, 1992

Authors' address:

G. P. Hellmann
Deutsches Kunststoff Institut
D-W-6100 Darmstadt, FRG

Progress in Colloid & Polymer Science   Progr Colloid Polym Sci 91:149–152 (1993)

Dynamic light scattering by non-ergodic media

J. G. H. Joosten

Department for Physical and Analytical Chemistry, DSM Research, Geleen, The Netherlands

Abstract: The basic notions of dynamic light scattering (DLS) on non-ergodic media are presented and discussed. We present the results, obtained by DLS, for the intermediate scattering function of polystyrene latex particles that are dispersed in aqueous polyacrylamide gels. It is shown that these systems display non-ergodic features, implying that the time-averaged intensity correlation function (ICF) is not equal to the ensemble-averaged ICF.

Key words: Light scattering – non-ergodic media – gels – tracer diffusion

The objective of this short communication is to highlight the basic notions of dynamic light scattering (DLS) by non-ergodic media. Details can be found in [1–3].

DLS measures the time-averaged time correlation function of the intensity $I(q, t)$ of light scattered by the sample in the direction described by the scattering vector $\mathbf{q}$, $|\mathbf{q}| = (4\pi/\lambda) \sin(\theta/2)$, λ and θ being the wavelength of the light and the scattering angle in the medium, respectively. This intensity is proportional to $|\rho(q, t)|^2$, where $\rho(q, t)$ is the qth spatial Fourier component of the fluctuations causing the scattering. The normalized density–density correlation function (intermediate scattering function, ISF) $f(q, \tau)$ is given by

$$f(q, \tau) = \frac{\langle \rho(q, 0)\rho^*(q, \tau) \rangle_E}{\langle |\rho(q)|^2 \rangle_E} \,, \qquad (1)$$

where τ denotes time and the subscript E indicates an ensemble average. A single DLS measurement provides an estimate of $g_T^{(2)}(q, \tau)$, the time-averaged normalized intensity correlation function (ICF):

$$g_T^{(2)}(q, \tau) \equiv \frac{\langle I(q, 0)I(q, \tau) \rangle_T}{\langle I(q, 0) \rangle_T^2} \,. \qquad (2)$$

where the intensity is given by $I(q, t) = |E(q, t)|^2$ and $\langle ... \rangle_T$ indicates a time average. For scattering processes in which the scattered field $E(q, \tau)$ is a zero-mean complex Gaussian variable, $g_E^{(2)}(q, \tau)$

can be expressed as [4]

$$g_E^{(2)}(q, \tau) = 1 + |\beta f(q, \tau)|^2 \,, \qquad (3)$$

where β is the spatial coherence factor which depends largely on the number of coherence areas seen by the detector [5]. In practice, data for $g_T^{(2)}(q, \tau)$ are taken from a correlator and by invoking the ergodicity theorem [6] the ensemble average is replaced by a time average. Therefore, in most practical situations, Eq. (3) (the Siegert relation) is used to find $f(q, \tau)$ from $g_T^{(2)}(q, \tau)$, but Pusey and van Megen [1] pointed out that for solid-like systems, like gels and colloidal glasses, Eq. (3) in conjunction with $g_E^{(2)}(q, \tau) = g_T^{(2)}(q, \tau)$ does not apply. The complications are caused by the fact that scatterers in those systems are restricted to particular regions of the sample and are only able to perform limited motions about fixed average positions. As a result, one particular sample of such a system is trapped in a restricted region of phase space defined by its average configuration and the extent of the fluctuations about this configuration. These solid-like systems may, thus, be regarded as a non-ergodic medium since a time-averaged measurement on a particular sample will not, in general, be equivalent to an ensemble-averaged measurement, i.e. one averaged over a representative number of all possible spatial configurations. By contrast, in liquid-like systems, the scatterers are able to diffuse throughout the sample so that, given enough time, the system evolves through a sequence of spatial configurations representative of the full ensemble of possible configurations (or, equivalently, of the whole of phase space). This implies that such a system during a single experiment can explore enough of phase space so that the time average, inherent in the measurement of a property, gives a good estimate of its ensemble average. Usually, the equivalence between time-averaged and ensemble-averaged time correlation functions is tacitly assumed. We anticipate, however, that next to gels [2, 3] and colloidal glasses [7, 8], for DLS experiments on systems such as those undergoing a sol–gel transition [9, 10] or those used for experiments on relaxation processes in amorphous polymers [11], the validity of this assumption has to be questioned.

Faced with a non-ergodic medium of interest one has two choices. The first is to construct the ensemble-averaged ICF by moving the sample through a series of positions so that different scattering volumes within the sample are illuminated. At each position an ICF of the scattered light is measured and finally the ensemble average is evaluated. After this tedious procedure, $f(q, \tau)$ can then be obtained in the usual way, i.e. using Eq. (3). Pusey and van Megen [1] worked out the alternative procedure, i.e. they calculated directly the scattering properties of a single scattering volume in the non-ergodic medium and used this in order to obtain $f(q, \tau)$ from a single ICF measurement.

The model starts off by noting that the total scattered field $E(q, \tau)$ in a non-ergodic medium is not a zero-mean complex Gaussian variable because of the spatial restrictions of the scatterers. Non-ergodicity of the medium is modelled by allowing only limited excursions of the scatterers about their fixed average positions. This means that the scattered field at the detector can be written as the sum of two components:

$$E(q, \tau) = E_F(q, \tau) + E_C(q) , \qquad (4)$$

where the fluctuating component $E_F(q, \tau)$ is a zero-mean complex Gaussian variable and $E_C(q)$ is a constant (time-independent) field. The constant component of the field depends explicitly on the relevant configuration of the scatterers and will, therefore, be different for different scattering volumes. In [1] it was found that $g_T^{(2)}(q, \tau)$ is given by

$$g_T^{(2)}(q, \tau) - 1 = Y^2 [\, f^2(q, \tau) - f^2(q, \infty)]$$
$$+ 2Y(1 - Y)[\, f(q, \tau) - f(q, \infty)] , \quad (5)$$

where

$$Y \equiv \frac{\langle I(q) \rangle_E}{\langle I(q) \rangle_T} \qquad (6)$$

and $f(q, \infty)$ is the limiting value of $f(q, \tau)$ for $\tau \to \infty$. One notes that for a fully fluctuating medium, i.e. an ergodic medium, for which $Y = 1$ and $f(q, \infty) = 0$, Eq. (5) reduces to Eq. (3). For a partially fluctuating medium, i.e. a non-ergodic medium, $f(q, \tau)$ starts at 1 and decays to a constant non-zero value, $f(q, \infty)$, which provides a measure of that fraction of the density (or strictly refractive index) fluctuations which is frozen in.

The quadratic equation (5) can be solved [2] for the intermediate scattering function giving

$$f(q, \tau) = 1 + \frac{1}{Y} [\, \sqrt{g_T^{(2)}(q, \tau) - \sigma_I^2} - 1] , \qquad (7)$$

where the mean-square intensity fluctuation σ_I^2 is given by

$$\sigma_I^2 \equiv \frac{\langle I^2(q) \rangle_T}{\langle I(q) \rangle_T^2} - 1 = g_T^{(2)}(q, 0) - 1 . \qquad (8)$$

Equation (7) applies to the scattered intensity at a single point in the far field. In practice, however, the exposed area of the detector is finite, and proper account must be taken of this partial spatial coherence resulting in $\beta < 1$. In [3] it is shown that for a detection system for which β is close to 1, the expression for $f(q, \tau)$ reads

$$f(q, \tau) = 1 + \frac{1}{Y}$$
$$\left[\sqrt{\frac{g_T^{(2)}(q, \tau) - 1 - \sigma_I^2}{\beta^2} + 1} - 1 \right]. \qquad (9)$$

Summarizing now the results, we conclude that $f(q, \tau)$ can be found from experimental data for $g_T^{(2)}(q, \tau)$, σ_I^2 and Y without making any assumptions regarding the functional form of $f(q, \tau)$. The parameter β can e.g. be determined from a DLS experiment using a standard dilute polystyrene latex dispersion in water. We emphasize that for DLS experiments on non-ergodic systems, the optical detection system should be designed such that $\beta \approx 1$, otherwise one gets a mixing of non-ergodicity effects with spatial integration effects [both effects reduce the contrast, σ_I^2, of $g_T^{(2)}(q, \tau)$].

We now present some results, obtained by DLS, for the diffusional motion of polystyrene latex spheres that are dispersed in a cross-linked polyacrylamide (PAA) gel. The preparation of the PAA gels has been described in [2]. Light scattering from these systems arises mainly from the tracer particles. It was shown [2] that these systems exhibit non-ergodic behavior, i.e. the time-averaged ICF, $g_T^{(2)}(q, \tau)$, is not equal to the ensemble-averaged ICF, $g_E^{(2)}(q, \tau)$. The non-ergodicity arises from the fact that the tracer particles (scatterers) are constrained to fixed average positions and are only

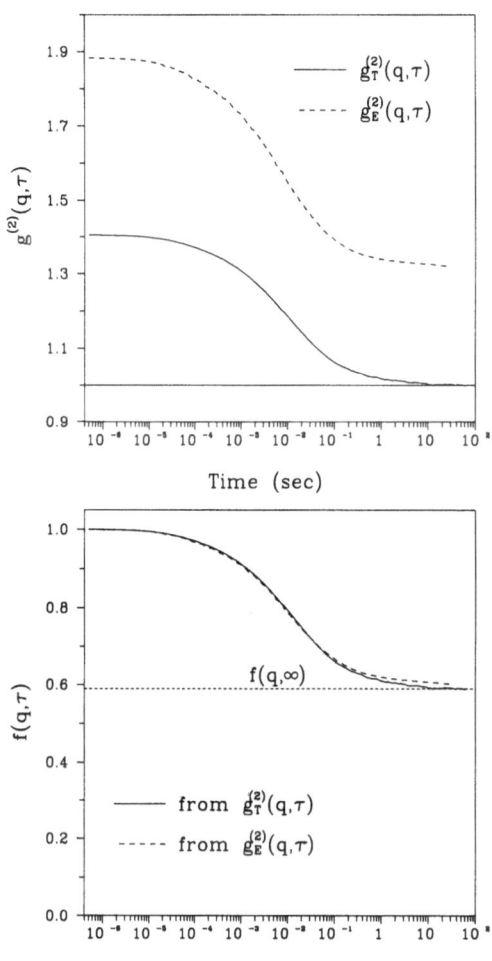

Fig. 1. Upper part: Ensemble-averaged ICF $g_E^{(2)}(q, \tau)$ (– – – –) and time-averaged ICF $g_T^{(2)}$ (————) obtained at a scattering angle $\theta = 90°$. Lower part: Intermediate scattering functions, $f(q, \tau)$, calculated from the results given in the upper part; (– – – –) $f(q, \tau)$ obtained from $g_E^{(2)}(q, \tau)$ using Eq. (3); (————) $f(q, \tau)$ obtained from $g_T^{(2)}(q, \tau)$ using Eq. (9); (· · · ·) $f(q, \infty)$ is the fraction of frozen-in fluctuations found from $g_T^{(2)}(q, \tau)$. Sample: An aqueous polyacrylamide gel cross-linked with bis-acrylamide containing 2.5 wt% polymer and 2.0×10^{-2} wt.% polystyrene spheres (diameter 82 nm); ratio acrylamide/bisacrylamide = 1.6 wt.%

able to perform limited Brownian motions about these average positions. The non-ergodic nature of the gel system is immediately apparent from the observation that by scanning through various positions in the sample a large variation in the time-averaged scattered intensity, $\langle I(q) \rangle_T$, is revealed. For example, for the sample given in Fig. 1, $\langle I(q) \rangle_T$ varies between $\approx 9 \times 10^4$ counts/s and $\approx 1.2 \times 10^6$ counts/s at a scattering angle $\theta = 90°$. The ensemble-averaged intensity, $\langle I(q) \rangle_E$, is measured to be 2.48×10^5 counts/s. An ensemble-averaged ICF, $g_E^{(2)}(q, \tau)$ (requiring a collection time of approximately 13 h), and a time-averaged ICF, $g_T^{(2)}(q, \tau)$ (requiring a measuring time of 30 min), are compared in Fig. 1. The ensemble-averaged ICF was constructed following the procedure described in [3], i.e. by adding unnormalized ICFs of 375 different speckles (= scattering volumes). Normalization of the final ICF is then accomplished by using the total number of photon counts and the total number of samples. From the $g_E^{(2)}(q, \tau)$ data, shown in Fig. 1, the spatial coherence factor β is found to be 0.94, whereas the value for β, as found from an ergodic system (a dilute aqueous latex dispersion), equals 0.945. This shows that the scattered field obeys Gaussian statistics when sampled over the full ensemble. The time-averaged ICF in Fig. 1 is measured on a speckle for which $\langle I(q) \rangle_T = 3.90 \times 10^5$ counts/s, resulting in $Y = 0.63$ [see Eq. (6)]. From $g_T^{(2)}(q, \tau)$ in Fig. 1, we calculate that $\sigma_I^2 = 0.404$ [see Eq. (8)]. The (normalized) baseline of $g_T^{(2)}(q, \tau)$ is calculated from the number of photon counts and the number of summations given by the correlator and as one observes from Fig. 1 the ICF decays properly to this baseline. Although the non-equivalence of $g_E^{(2)}(q, \tau)$ and $g_T^{(2)}(q, \tau)$ is obvious from Fig. 1, the intermediate scattering functions, $f(q, \tau)$, obtained from the two ICFs should be equivalent. Obviously, $f(q, \tau)$ can be calculated from $g_E^{(2)}(q, \tau)$ by using Eq. (3). To obtain $f(q, \tau)$ from $g_T^{(2)}(q, \tau)$, we used Eq. (9), thereby taking $\beta = 0.945$, $Y = 0.63$ and $\sigma_I^2 = 0.404$. The results for both $f(q, \tau)$'s are also given in Fig. 1 (lower part). As can be seen from Fig. 1 the agreement between the two $f(q, \tau)$'s is very good, implying that the theory given by Pusey and van Megen [1] can be used to calculate $f(q, \tau)$ from $g_T^{(2)}(q, \tau)$. Moreover, it was shown in [2] that this procedure yields intermediate scattering functions that are independent of the intensity, $\langle I(q) \rangle_T$, of a particular speckle. It is obvious that for analyzing the experimental ISF it is still necessary to construct a theoretical model that describes the functional form of $f(q, \tau)$ (or certain features of it like the first cumulant or the frozen-in fraction $f(q, \infty)$), but this is beyond the scope of this paper (see, however, [2, 3, 7, 8]).

Acknowledgement

I thank Prof. Peter Pusey for fruitful discussions and Dr. Jennifer McCarthy for help in the experiments.

References

1. Pusey PN, Megen W van (1989) Physica A 157:705
2. Joosten JGH, Geladé ETF, Pusey PN (1990) Phys Rev A 42:2161
3. Joosten JGH, McCarthy JL, Pusey PN (1991) Macromolecules 24:6690
4. Pusey PN (1977) In: Cummins HZ, Pike ER (eds) Photon Correlation Spectroscopy and Velocimetry. Plenum Press, New York
5. Jakeman E (1974) In: Cummins HZ, Pike ER (eds) Photon Correlation and Light Beating Spectroscopy. Plenum Press, New York
6. Berne BJ, Pecora R (1976) Dynamic Light Scattering. Wiley, New York
7. Pusey PN, Megen W van (1987) Phys Rev Lett 59:2083
8. Megen W van, Pusey PN (1991) Phys Rev A 43:5429
9. Martin JE, Wilcoxon J (1989) Phys Rev A 39:252
10. Kaufman VR, Müller SC, Müller KH, Avnir D (1987) Chem Phys Lett 142:551
11. Fytas G (1989) Macromolecules 22:211

Author's address:

Dr. Jacques G. H. Joosten
Department for Physical & Analytical Chemistry
DSM Research
P.O. Box 18
6160 MD Geleen, The Netherlands

Diffusion of flexible and semirigid polymers confined to the pore spaces in porous glass

K. H. Langley[1]), I. Teraoka[2]), and F. E. Karasz[2])

[1]) Department of Physics & Astronomy
[2]) Polymer Science and Engineering Department, University of Massachusetts, Amherst, MA, USA

Abstract: The dynamics of polymer chains in solution within the pore spaces of controlled pore glasses has been studied experimentally by using dynamic light scattering and theoretically by computer simulation. For flexible chains, we have explored the dependence of diffusion on the degree of confinement (i.e. on molecular weight or on R_h/R_p where R_h is the polymer hydrodynamic radius in free solution and R_p is the glass pore radius), molecular architecture, and time or distance scale of the diffusion measurement. Flexible chain diffusion over macroscopic distances may be understood in terms of hydrodynamic interactions with the pore wall at low and intermediate confinement, and in terms of an entropic barrier model at strong confinement. At intermediate scattering wavevector ($qR_p \simeq 1$) the apparent diffusion coefficient is time-dependent, and exhibits a crossover from the free-solution value at short times to a smaller macroscopic (hindered) diffusion coefficient value at later times. Semirigid chains experience much stronger hindrance than flexible chains. We have developed a model of the dynamics based on reptation theory. The mean squared displacement of a sufficiently long chain is expected to display a power law time dependence; the exponent depends on the time, changing from approximately 1/4 at short times to 1 at long times. Diffusion measurements of a series of poly(hexylisocyanate) polymers in porous silica display a definite crossover to much slower hindered diffusion at higher molecular weight.

Key words: Diffusion – porous glass – semirigid polymers – dynamic light scattering

In a series of recent papers [1–4], we have described dynamic light-scattering studies of the diffusion of a flexible polymer (polystyrene) in a good solvent (2-fluorotoluene) confined in the pores of controlled-pore glasses. The diffusion coefficient was found to depend on the distance and time scale on which the measurement is made. As the distance scale of the measurement is increased to a length greater than the pore dimensions ($qR_p < \sim 0.7$, where R_p is the pore radius), the measurement averages over details of the pore structure, and D asymptotically approaches what is referred to as the macroscopic diffusion coefficient. At intermediate distance scales ($qR_p \simeq 1$) the initial diffusion coefficient is nearly identical to that in free solution, but then at later times crosses over to a smaller value which appears to be equal to the macroscopic diffusion coefficient. For polymers which are small compared with the pore size ($R_h/R_p < \sim 0.8$, where R_h is the polymer hydrodynamic radius) the ratio of the macroscopic diffusion coefficient to the diffusion coefficient in free solution, D/D_0, depends on R_h/R_p in a manner that agrees well with theories of hydrodynamic drag due to the presence of the pore walls. For strongly confined polymers ($R_h/R_p > \sim 1$), D/D_0 is consistent with a model which considers the entropic barriers a polymer must overcome in order to move through small openings which connect regions of less strict confinement. We also showed that the macroscopic diffusion rate of a star polymer is less than that of a linear polymer of the same hydrodynamic radius as one would expect.

The diffusion of semirigid polymers, in contrast to flexible polymers as discussed above, is dominated by quite different mechanisms if the persistence length of the polymer is comparable to, or larger than the pore radius. A long semirigid chain is forced into an extended conformation by the unyielding pore walls. We may think of the path of the polymer chain as a spatial curve wiggling about the centerline of a tortuous tube which is characterized by a diameter and a persistence length determined by the pore structure of the rigid glass (Fig. 1). This picture, with its suggestion of a primitive path through fixed obstacles, has prompted us to formulate a model of the diffusion based on reptation theory [5]. The principal results of our theory, which will be described in detail elsewhere, are summarized as follows. Concerning the concentration of chains inside the pore space (partitioning): calculation and computer simulation show that if we compare chains of the same end-to-end distance in the pore network and identical concentration in the solution surrounding the porous material, the highest concentration occurs for chains with a free-solution persistence length which lies in the range between the pore radius and the persistence length of the tube or primitive path. Thus, it is this region which is experimentally most accessible. Motion of the center of mass of the primitive path is restricted to be along the pore, and obeys the assumptions of

the Doi and Edwards reptation theory [5] for entangled polymers, although the conformation of the primitive path is different. We can obtain the long-time diffusion coefficient as a function of chain and pore geometry. At short times the normal mode analysis of Aragon and Pecora [6] describes chain dynamics in the pore network. Combining the normal mode and reptation analysis, we find that for a sufficiently long, semirigid polymer chain, the mean squared monomer displacement proceeds at a rate proportional to t^{α}. α is approximately 1/4 for the short-time motion inside the pore, $\alpha \simeq 1$ for translation along a single-pore branch, and $\alpha = 1$ for long-time diffusion over macroscopic distances which average over many primitive paths. An additional intermediate range characterized by $\alpha \simeq 1/2$ exists, in principle, but may not be observable.

We have used dynamic light scattering to measure the long-time macroscopic diffusion coefficient of a semirigid chain, poly(hexylisocyanate), in solution in 2-fluorotoluene in the pore space of silica controlled-pore glass. The results (shown in Fig. 2) are for a series of molecular weights ranging from 1.03×10^5 to 3.82×10^5. The upper curve is the diffusion coefficient in free solution and the lower curve is the hindered diffusion coefficient measured inside the porous glass. At the minimum molecular weight, the ratio of contour length to persistence length L_c/L_p is approximately 4.0 (a

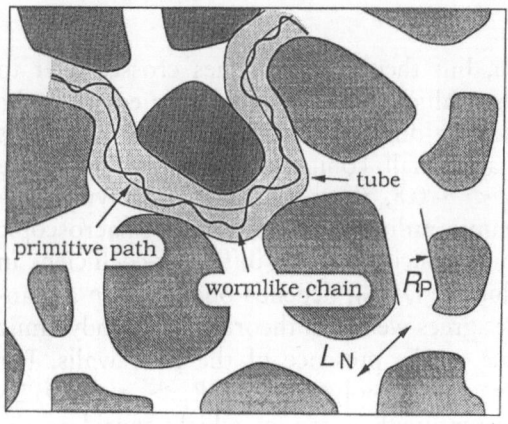

Fig. 1. A semirigid chain diffuses in a tube-like space (light shading) between the solid-phase regions (dark shading) of the porous glass

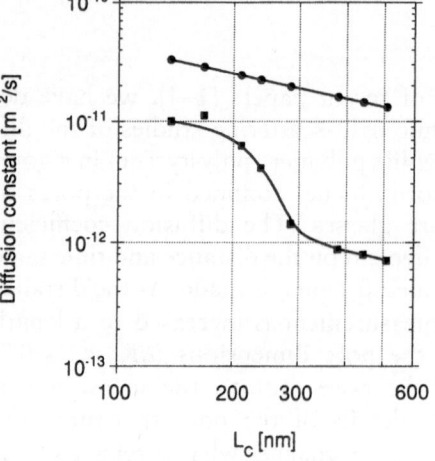

Fig. 2. Diffusion coefficient of semirigid poly(isocyanate) in free solution (solid circles) and in porous glass (squares) vs. polymer contour length (L_c)

relatively stiff polymer) and in all cases the persistence length (34 nm) is somewhat greater than the pore radius (27.5 nm). At lower contour length, diffusion in the pores, as expected, is much slower (by a factor of about 3) than in free solution. This is due primarily to hydrodynamic interaction with the pore walls, but is also partially due to steric interference between semirigid polymer and the pore walls. The sudden decrease in D at higher molecular weight probably corresponds to the point at which the average polymer end-to-end distance becomes comparable to the mean length of the pore cylindrical sections, although the pore structure is not well-known enough to confirm this.

Acknowledgement

This work was supported in part by the Air Force Office of Scientific Research Grant No. 91-001.

References

1. Biship MT, Langley KH, Karasz FE (1989) Macromolecules 22:1220–1231
2. Easwar N, Langley KH, Karasz FE (1990) Macromolecules 23:738–745
3. Guo Y, Langley KH, Karasz FE (1990) Macromolecules 23:2022–2027
4. Guo Y, Langley KH, Karasz FE (1990) J Chem Phys 93:7457–7462
5. Doi M, Edwards SF (1986) The Theory of Polymer Dynamics. Clarendon, Oxford
6. Aragon SR, Pecora R (1985) Macromolecules 18:1868–1875

Authors' address:

Kenneth H. Langley
Department of Physics & Astronomy
University of Massachusetts
Amherst, MA 01003, USA

Coils, globules and solubilization of a thermosensitive polymer

J. Rička, M. Meewes, Ch. Quellet, and Th. Binkert

Institute of Applied Physics, University of Bern, Switzerland

Abstract: The thermosensitive ternary system poly(N-isopropylacrylamide)-SDS-water exhibits a remarkable coil-globule transition whose features are governed by polymer-surfactant interactions. We report on light scattering studies of the collapse and expansion of the polymer chain as well as preliminary results (obtained with time resolved fluorescence depolarization) on the formation and rotational dynamics of polymer induced surfactant micelles.

Key words: Coil-globule-transition – polymer-surfactant interactions – poly(N-isopropylacrylamide)

The ternary system consisting of the thermosensitive polymer poly(N-isopropylacrylamide) (PNIPAM), the surfactant sodium dodecylsulphate (SDS) and the solvent water combines in an intriguing way several important phenomena of colloid and polymer science. They include a temperature-induced coil–globule transition, cooperative association of the surfactant with the polymer and two forms of solubilization of the polymer. For the investigation of this interesting system, we have employed two variants of time-correlated photon counting, each working on a different length and time scale. Dynamic and static light scattering provided information on the dynamics and conformation on the scale of whole macromolecules, whereas time-resolved fluorescence depolarisation elucidated the sub-microscopic features on the scale of a few polymer segments. From our measurements, the following picture of the PNIPAM–SDS–water system is emerging.

The polymer is soluble (in the form of coils) in cold pure water but exhibits a lower consolute temperature LCST of 34 °C. A macroscopical manifestation of the phase transition at 34 °C is precipitation of the polymer. Microscopically, we observe [1] (with time-resolved fluorescence depolarization on a nanosecond time scale) at LCST an abrupt "freezing" of reorientational motions of fluorescent labels covalently bound to the polymer

backbone. This indicates that the temperature-induced conformational changes result in a collapse of polymer coils into a dense state and this in turn invokes the notion of a coil–globule transition. Since, however, the same attractive interactions which are responsible for the collapse of the coils are acting also between different molecules, the coil–globule transition is usually masked by aggregation. In our system the aggregation has been a very serious problem prohibiting a conclusive interpretation of light scattering data. Using a sensitive enough technique, we have observed aggregation above 34 °C down to a polymer concentration of only 0.01 ppm.

Aggregation can, however, be prevented by the addition of a surfactant [2]. Only 10 mg/l SDS are sufficient to clarify the turbid polymer solutions above LCST. At 250 mg/l the aggregates are completely dispersed (intermolecular solubilization) and one can study the behaviour of isolated polymer molecules in the whole temperature range [3]. Upon heating, the polymer undergoes indeed a coil-to-globule phase transition with a volume reduction by a factor of more than 300 for a polymer molecular weight of 7×10^6 (see Fig. 1). In dynamic light scattering the collapse manifests itself in a decrease of the hydrodynamic radius as well as in the vanishing of the internal motion contribution to the correlation function. With

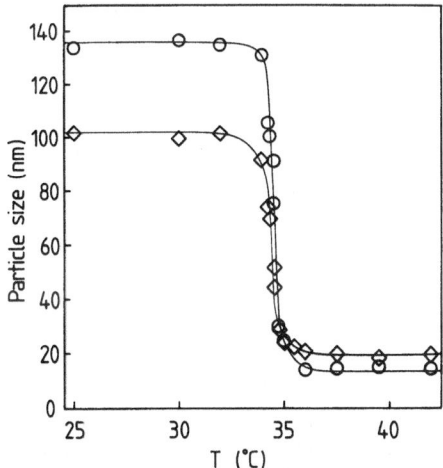

Fig. 1. Hydrodynamic radius ($\diamond$) and radius of gyration ($\bigcirc$) as a function of temperature. Surfactant concentration is 300 mg/l

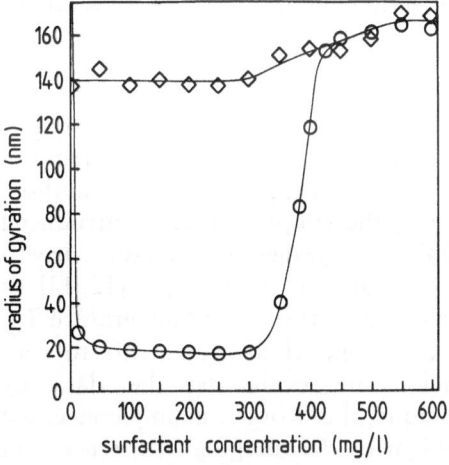

Fig. 2. Radius of gyration as a function of surfactant concentration at temperatures below and above the LCST: ($\diamond$) 25 °C; ($\bigcirc$) 36 °C

static light scattering we observe corresponding changes in the structure factor and radius of gyration.

The transition temperature of the collapse depends on the SDS concentration. It is first constant and equal to the LCST, but then begins to increase above 300 mg/l surfactant. Therefore, when in-creasing the surfactant concentration at constant temperatures above the LCST, we cross the phase boundary and observe intramolecular solubilization, i.e. a surfactant-induced globule-to-coil transition (see Fig. 2). Below the LCST, the surfactant causes an expansion of the polymer coils also setting on at 300 mg/l (Fig. 2).

It has been suggested that at 300 mg/l a critical adsorption concentration (CAC) is reached, above which the surfactant binds cooperatively, in the form of micelles, to the polymer. Currently, we are investigating the surfactant binding by monitoring the reorientational relaxation of an amphiphilic fluorescent probe pyrene–$C_{11}COOH$. Our preliminary results indicate that in the presence of the polymer, micelles begin to form indeed already at a CAC of approximately 300 mg/l, far below the critical micellization concentration (CMC) of the surfactant in pure water. Below LCST we observe two reorientational modes with contributions depending on the SDS content. A slower mode dominates in the vicinity of CAC. With increasing SDS concentration a faster mode appears and dominates eventually the relaxation behaviour when CMC is approached. The relaxation time of the slower mode is consistent with micelles firmly bound to the polymer chain, whereas the faster mode corresponds to free micelles in the coil region. Both modes do freeze above the transition temperature.

References

1. Binkert Th, Oberreich J, Meewes M, Nyffenegger R, Rička J (1991) Macromolecules 24:5806
2. Rička J, Meewes M, Nyffenegger R, Binkert Th (1990) Phys Rev Lett 65:657
3. Meewes M, Rička J, de Silva M, Nyffenegger R, Binkert Th (1991) Macromolecules 24:5811

Authors' address:

J. Rička
Universität Bern
Institut für Angewandte Physik
Sidlerstr. 5
CH-3012 Bern
Switzerland

Progress in Colloid & Polymer Science   Progr Colloid Polym Sci 91:158–161 (1993)

Relation between main- and normal-mode relaxation.
A dielectric study on poly(propyleneoxide)

E. Schlosser and A. Schönhals

Zentrum für Makromolekulare Chemie, Berlin, FRG

Abstract: The relation between main- and normal-mode relaxation is studied on poly(propyleneoxide) of different molecular weights by dielectric measurements in a very large frequency range. The evaluation by a special technique shows a merging of α-relaxation and of normal-mode relaxation at an essential higher temperature than the Vogel temperature of the α-relaxation. This means the Rouse model is not applicable to the normal-mode relaxation near the glass transition temperature.

Key words: α-relaxation – normal-mode relaxation – dielectric measurements – Arrhenius plot – entanglements

Introduction

Molecular motions in polymers are controlled by very different length scales ranging from the local segmental motion, responsible for the glass transition to cooperative motions involving the whole polymer chain known as normal-mode relaxation. The relation between segmental-mode (α) and normal-mode (n) relaxation in their dependencies on temperature has been studied by several authors before [1–9], mostly by dielectric measurements. Generally, it has been found that the dependence of the loss peak frequency f_p on temperature T for the segmental mode ($f_{p\alpha}$) as well as for the normal mode (f_{pn}) obeys the Vogel–Fulcher (VF) equation [10, 11]

$$\log f_{pi} = \log f_{\infty i} - A_i/(T - T_{0i}) , \qquad (1)$$

where $f_{\infty i}$, A_i are constants and T_{0i} is the so-called Vogel temperature ($i = \alpha$, n). Further, the higher the molecular weight M of the polymer, the greater is the ratio $f_{p\alpha}/f_{pn}$. These facts have been explained by the application of the Rouse theory and the concept of molecular friction coefficient to the normal-mode process [1, 2, 7, 9] predicting $T_{0n} = T_{0\alpha}$. But on account of the limiting frequency range and the difficulty of resolving the contributions from two overlapping mechanisms,

it is not clear up to now if the normal-mode and the α-relaxation merges at $T_{0\alpha}$ or at a higher temperature ($T_{0n} < T_{0\alpha}$) as suggested by some authors [5–8]. Moreover, the coupling model, introduced by Ngai, predicts a stronger temperature dependence of $f_{p\alpha}$ compared to that of f_{pn} [12, 13] as T approaches the glass transition temperature T_g.

Extending the range of measurements to very low frequencies and evaluating the data by a special technique, this work is mainly concerned with the problem of the merging of α-relaxation and of normal-mode process. To observe the normal-mode relaxation by dielectric measurements poly(propyleneoxide) (PPO) was chosen, which has a dipole component parallel to its chain backbone. Three samples of PPO having different molecular weights were measured: PPO-1, PPO-2, and PPO-4 with M (g/mol) = 1000, 2000, and 4000, respectively.

Measurement and evaluation

For measurement the sample was filled into a guard ring condenser of silver-coated brass electrodes, contained in a glass vessel to protect the sample against humidity, which was placed in an air-stream cryostat (temperature stability

±0.1 K). The isothermal dielectric behavior characterized by the complex dielectric permittivity

$$\varepsilon^*(f) = \varepsilon'(f) - i\varepsilon''(f) , \qquad (2)$$

where f is the frequency, ε' the real part, ε'' the imaginary part and $i = \sqrt{-1}$ was measured over 10 decades in frequency by several equipments. From 10^{-5} to 1 Hz a time-domain spectrometer was employed [14] and the complex permittivity was calculated by a special evaluation strategy [15]. From 1 Hz up to 10^5 Hz the measurement was carried out directly by ac-bridges [16, 17]. Figure 1 shows $\log \varepsilon''$ vs. $\log f$ for PPO-4 measured at $T = 228$ K. Clearly, two relaxation processes indicated by the peaks in ε'' can be seen, where the high-frequency peak is due to the α-relaxation and the low-frequency one is known as normal-mode relaxation.

To separate these two processes and to determine f_{pi} also in the case of strong overlapping of the α-relaxation and the normal-mode relaxation an evaluation strategy [15] based on the model function introduced by Havriliak and Negami [18] (HN-function) was applied. The HN-function reads

$$\varepsilon^*(f) - \varepsilon_\infty = \Delta\varepsilon[1 + (if/f_0)^\beta]^{-\gamma} \qquad (3)$$

with the HN-parameters $\Delta\varepsilon$ being the intensity, f_0 the characteristic frequency, which is nearly equal to the peak frequency f_p, β, γ the shape parameters, and $\varepsilon_\infty = \varepsilon'(f)$ for $f \gg f_p$. The HN-parameters for

each process were determined by simultaneously fitting two HN-functions to the experimental data. This procedure enables

1) the separation of n- and α-relaxation, also in the case of strong overlapping (cf. Fig. 1),
2) a consistent evaluation of the data obtained in the frequency and/or time domain,
3) the separation of neighboring processes, e.g. conductivity.

Results and discussion

The temperature dependence of the peak position for both the α- and the n-relaxation obeys the VF-equation [Eq. (1)] with the parameters given in Table 1. As shown in Fig. 2 (Arrhenius plot), the

Table 1. VF-parameters

	α-process	Normal-mode process		
	PPO-1, 2, 4	PPO-1	PPO-2	PPO-4
T_0 (K)	166	15	156	153
$\log(f_{p\infty})$ (Hz)	11.8	11.6	10.5	10.2
A (K)	459	651	582	643

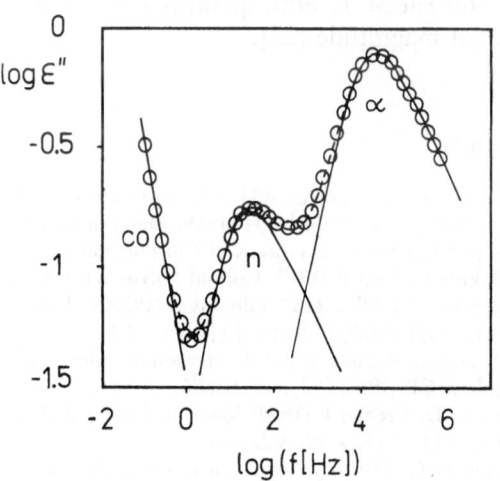

Fig. 1. Decomposition of the measured ε''–f dependence into n-relaxation, α-relaxation and conductivity (co) process (PPO-4, $T = 228$ K)

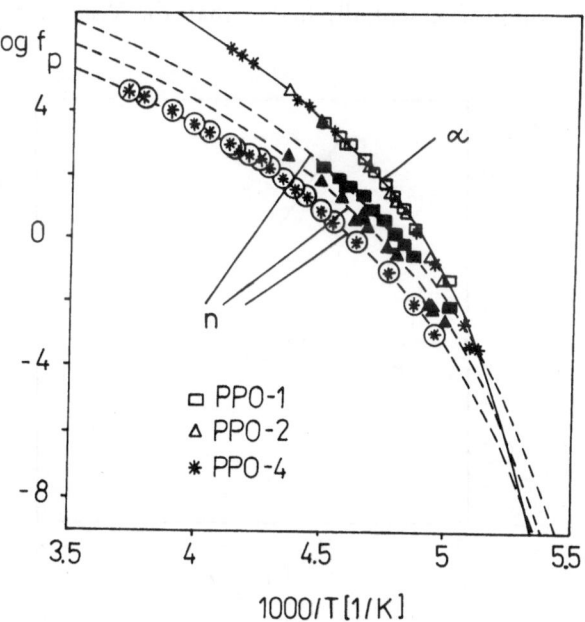

Fig. 2. Arrhenius plot of n- and α-relaxation for PPO-1, 2, 4

α-relaxation can be described by the same set of VF-parameters. This observation suggests that our samples should have a molecular weight M greater than the critical molecular weight M_c, which is characteristic for entanglement. However, it is known by previous experience that for flexible polymers, M_c has an order of magnitude of 10 000 [9], which is significantly greater than the molecular weight of the samples employed. Therefore, the coincidence of the VF-parameters cannot be caused by normal entanglement. But it may be an argument for the existence of transient entanglements [19–21] in PPO caused by hydrogen bonds of OH-end-groups.

Contrary to the α-process, the peak frequency f_{pn} depends on molecular weight according to the formula

$$f_{pn} \sim M^{-A} , \tag{4}$$

where $A = 2$ is predicted by the Rouse theory and $A = 3.4–4$ by the reptation model [7, 9]. As shown in Fig. 3 (e.g. for $T = 222$ K) the slope of $\log f_p$ versus $\log M$ provides $A \approx 3$, which is a further argument for transient entangled chains.

From extensive dielectric measurements down to very low frequencies, the VF-parameters of Table 1 should be sufficiently accurate to permit the extrapolation drawn by dashed lines in Fig. 2, which shows clearly the intersection of the VF-curves for α-relaxation and the n-relaxation at a temperature T_s which is essentially higher than the Vogel temperature $T_{0\alpha}$ of the α-relaxation, suggesting that the application of the Rouse theory to normal-mode relaxation breaks down near T_g. Also a detailed dielectric study of oligomeric polyisoprene displays the same behavior [22].

An analysis of these experimental facts in the framework of the coupling model given in ref. [23] reproduces quantitatively the observed behavior. To enlighten these results from a more microscopic point of view we start from the conception that the α-relaxation is a cooperative process which can be characterized by a correlation length ξ [24–27]. Within the framework of this idea the strong decrease of $f_{p\alpha}$ with decreasing temperature is explained by the strong temperature dependence of the correlation length $\xi(T)$ which goes to infinity as T approaches $T_{0\alpha}$. On the other side, the normal-mode process is governed by the fluctuation of the end-to-end vector $\sqrt{\langle R^2 \rangle}$ [7, 9], which increases with the molecular weight but changes scarcely with temperature.

The merging of both processes should take place if by lowering the temperature the correlation length (for the α-relaxation) is increased to the order of magnitude of the end-to-end distance

$$\xi(T_s) \approx \sqrt{\langle R^2 \rangle} . \tag{5}$$

Since $\langle R^2 \rangle$ increases with M, the merging of both processes should shift to lower temperatures for higher molecular weights, as it can be seen in Fig. 2. Also a quantitative comparison of ξ using the fluctuation approach [25] and $\sqrt{\langle R^2 \rangle}$, shows that for polyisoprene at T_g both quantities are of the same order of magnitude [22].

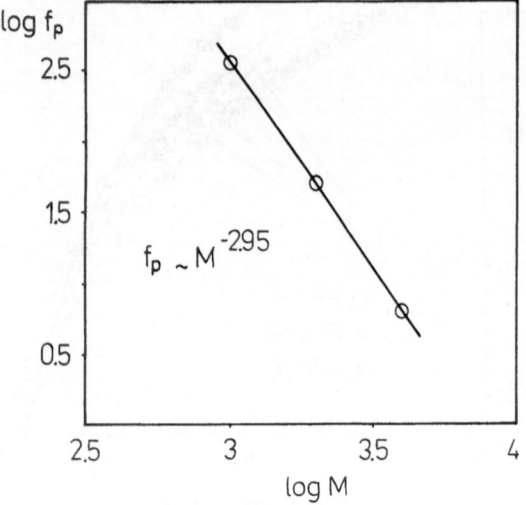

Fig. 3. Molecular weight dependence of the peak position of the n-process for $T = 222$ K

References

1. Baur ME, Stockmayer WH (1965) J Chem Phys 43:4319
2. Stockmayer WH, Burke JJ (1969) Macromolecules 2:647
3. Alper T, Barlow A, Gray R (1976) Polymer 17:665
4. Bakule R, Stoll B (1977) Colloid Polym Sci 255:1176
5. Beevers MS, Elliot DA, Williams G (1979) Polymer 20:785
6. Plazek DJ (1980) Polymer J (Japan) 12:43
7. Adachi K, Kotaka T (1984) Macromolecules 17:120
8. Johari GP (1986) Polymer 27:866
9. Boese D, Kremer F (1990) Macromolecules 23:829
10. Vogel H (1921) Z Phys 22:645
11. Fulcher GS (1925) J Amer Chem Sco 8:339, 789
12. Ngai KL, Plazek PJ (1986) J Polym Sci Polym Phys Ed 24:619
13. Ngai KL, Plazek DJ, Deo SS (1987) Macromolecules 20:3047

14. Schlosser E, Schönhals A (1991) Polymer 32:2135
15. Schlosser E, Schönhals A (1989) Colloid Polym Sci 267:963
16. Schlosser E (1962) Plaste Kautschuk 9:582
17. Schlosser E, Horn G (1963) Z Exp Techn Physik 11:145
18. Havriliak S, Negami S (1966) J Polym Sci, Polym Symp 14:89
19. Heinrich G, Alig I, Donth E (1988) Polymer 29:1198
20. Alig I, Donth E, Schenk W, Hörig S, Wohlfarth Ch (1988) Polymer 29:2081
21. Fleicher G, Helmstedt M, Alig I (1990) Polym Comm 31:409
22. Schönhals A (1992) Macromolecules, to be submitted
23. Ngai KL, Schönhals A, Schlosser E (1992) Macromolecules, in press
24. Adam G, Gibbs JH (1965) J Chem Phys 43:139
25. Donth E (1981) Glasübergang. Akademie-Verlag, Berlin
26. Fischer EW, Donth E, Steffen W (1992) Phys Rev Lett 68:2344
27. Sillescu H (1991) Application of Scattering Methods to the Dynamics of Polymer Systems, 27th Europhysics Conf, Macromolecular Phys, Heraklion, Crete, Greece

Received January 20, 1992;
accepted June 15, 1992

Authors' address:

Dr. E. Schlosser
Zentrum für Makromolekulare Chemie
Rudower Chaussee 5
D-O-1199 Berlin, FRG

Progress in Colloid & Polymer Science

Progr Colloid Polym Sci 91:162–164 (1993)

Structural relaxation in a low molecular weight poly(methylphenyl siloxane)

F. Stieber[1]), G. Floudas[2]), I. Alig[3]), and G. Fytas[1])

[1]) Foundation for Research and Technology-Hellas, Heraklion, Crete, Greece
[2]) Max-Planck-Institut für Polymerforschung, Mainz, FRG
[3]) Inst. Phys. Chem. University Köln, Köln, FRG

Abstract: The results of dynamic light scattering, ultrasonic and dielectric measurements of a bulk poly(methylphenyl siloxane) with $M_W = 1800$ g/mol over a broad temperature and frequency range are reported. The temperature dependence of the reorientational relaxation times exhibit very fragile behaviour and indicates the decoupling of density and orientational relaxation times at temperatures higher than $1.2 T_g$. This suggests a major contribution of volume structural rearrangements to longitudinal relaxation processes for this sample in the ultrasonic and hypersonic frequency region.

Key words: Reorientational relaxation – longitudinal relaxation – poly(methyl-phenyl siloxane)

Introduction

In a series of papers [1–4], investigations on local dynamics of bulk poly(methylphenyl siloxane) (PMPS) – a highly flexible polymer – were reported. A compilation of all available relaxation data on this polymer is presented in Ref. [4]. It becomes evident, that up to 70 K above the glass transition temperature T_g the temperature (T) dependence of the reorientational and structural relaxation times for different samples follows the same tamperature dependence, thus implying strong coupling between these processes. At $T - T_g > 70$ K, Brillouin measurements yield by an order of magnitude shorter structural relaxation times than the reorientational times. This suggests that in this T-range molecular orientational processes do not dominate the structural rearrangements. This behaviour is more pronounced in molecular glass formers [5]. In the present report, we complement dynamic light-scattering techniques with dielectric and ultrasonic measurements performed on the same low molecular weight PMPS sample in order to compare reorientational and structural dynamics over a wide T-range.

Experimental

Polarized Brillouin spectra (BS) were taken with a piezoelectrically scanned, cavity stabilized single-pass Fabry–Perot interferometer (FPI). The excitation source was an argon-ion single-mode laser, operating at 488 nm. BS spectra were obtained at three scattering angles θ (30°, 45° and 90°) over the temperature range 260–440 K. The scattering vector $q = (4\pi\lambda^{-1}n)\sin(\theta/2)$ was calculated using $n = 1.641 - 3.6 \times 10^{-4} T$ (T in K). The Brillouin shift ω_B and line width Γ_B were extracted from the BS spectra by a fitting procedure, taking into account the overlap between the Rayleigh and Brillouin peaks. Depolarized Rayleigh spectra (DRS) far above T_g were obtained at a scattering angle of 90° and temperatures of 353, 393 and 413 K. The line widths were calculated by a non-Lorentzian fitting procedure [6]. At low T, near T_g, the correlation function $C_{VH}(t)$ of the depolarized light scattering was measured with a Malvern (K7027) log-lin clipped correlator in the temperature range 238–245 K. The correlation times τ_0 were calculated from the fit of the stretched exponential function $[\exp(-(t/\tau_0)^\beta]$ (where β is a measure of the distribution of relaxation times) to the experimental $C_{VH}(t)$. The ultrasonic measurements (US) were performed with a conventional pulse-transmission technique at a fixed frequency of 1 and 8.4 MHz, using a HP-storage oscilloscope DSO 54501A, in a T-range from 170 to 410 K and yield the sound velocity u and absorption α with an accuracy of 3% and 10%, respectively. The dielectric measurements (DS) were made in the frequency range $10–10^7$ Hz with a HP 4192A impedance analyser at temperatures from 227 to 290 K. The PMPS sample has a weight-averaged molecular weight $M_S = 1800$ and a T_g of 233 K obtained at a heating rate of 10 K/min.

Results

To compare ultrasonic and hypersonic experiments with reorientational times obtained from PCS, DS and DRS measurements, we have consistently used the retardation time obtained from the loss compliance D'' data. The latter is defined [7] from the measurable quantities: α, u, ω_B and Γ_B, through $D'' = M''/[(M')^2 + (M'')^2]$ (M is the longitudinal modulus), when $\alpha u/\omega \ll 1$ as

$$D'' \simeq \frac{2\alpha}{\omega_B \rho u} = \frac{2\Gamma_B q^2}{\rho \omega_B^3}. \qquad (1)$$

The T-dependence of density is given by $\rho = [1115 - 0.8(T - 293)]\,\mathrm{K\,gm^{-3}}$. Figure 1 shows a plot of D'' vs. T. The higher values of D'' obtained from BS are due to an excess line width $\Gamma_{B,\,exc}$ which arises from an additional internal relaxation mechanism associated with the phenyl group. $\Gamma_{B,\,exc}$, being proportional to q^2, becomes less important at low scattering angles. The larger breadth of D'' at hypersonic frequencies and, hence, at higher T as compared to the ultrasonic D'' reflects the decrease in the rate of change of τ with T as the temperature increases (Fig. 2 below). This also accounts for the large disparity between the temperatures at maximum M'' and D'' at higher T. This difference is shown in Fig. 2, which depicts the variation of the longitudinal relaxation times τ_L and of the reorientational relaxation times τ_{or} vs. T_g/T. The solid line denotes a fit of the Vogel–Fulcher–Tammann–Hesse (VFTH) equation to the reorientation times with parameters: $\log(\tau/s) = -10.8$, $B = 420$ K and $T_0 = 218$ K. These parameters for the low molecular weight PMPS exhibit quite fragile behaviour indicated by the unusually low fragility factor [8], $D = B/T_0 = 1.9$. However, as anticipated for low D values, the VFTH fit shows systematic deviations at $T > 1.2\,T_g$ [9] as indicated by the orientational times for two different molecular weight PMPS samples ($M_W = 28\,500$, $T_g = 247$ K and $M_W = 1800$, $T_g = 233$ K) in Fig. 2.

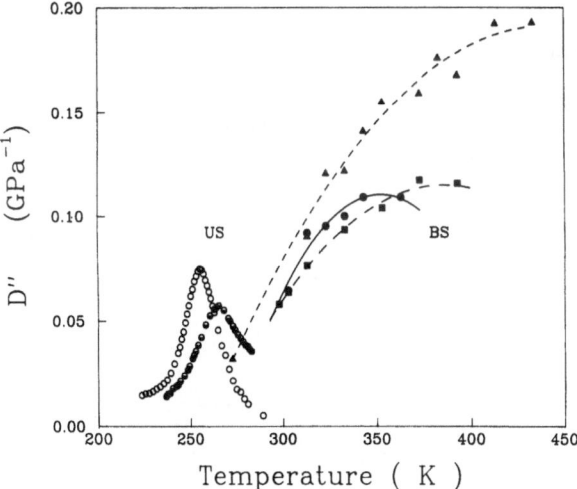

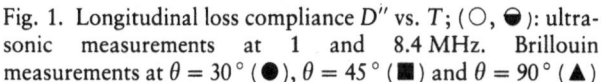

Fig. 1. Longitudinal loss compliance D'' vs. T; ($\bigcirc$, $\ominus$): ultrasonic measurements at 1 and 8.4 MHz. Brillouin measurements at $\theta = 30°$ ($\bullet$), $\theta = 45°$ ($\blacksquare$) and $\theta = 90°$ ($\blacktriangle$)

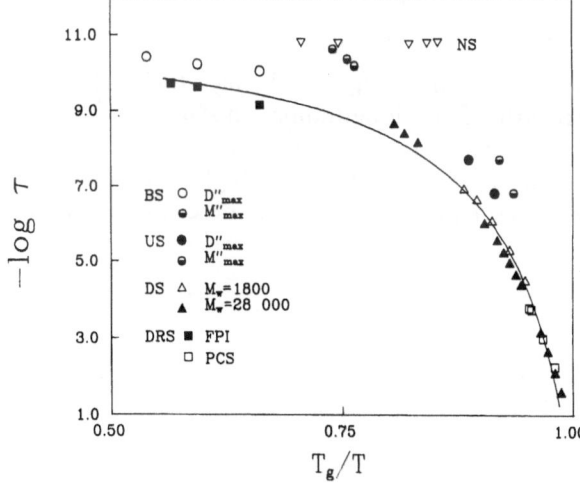

Fig. 2. Temperature dependence of the reorientational relaxation times ($\blacktriangle$: from Ref. [2]) the longitudinal relaxation times and the neutron scattering (NS) data from Ref. [4]

Discussion

Several trends are apparent in Fig. 2. The relaxation times obtained from BS are by one order of magnitude shorter than the reorientational times obtained from FPI-DRS. Comparing BS with neutron scattering data it is evident that localized "fast" processes, associated with phenylene flips [4] are responsible for the relaxation processes measured in BS [10]. At high T, rotational dynamics probed by FPI-DRS arise from cooperative segmental orientational motion requiring a larger volume and, in addition, display non-exponential decay [6]. In contrast to the high molecular weight sample ($M_W = 28\,500$) the relaxation times obtained from US for the small molecular weight PMPS sample ($M_W = 1800$) are also shorter than the reorientation time. This behaviour can be rationalized by the fact that for low molecular weights and T far above T_g the loss M'' is dominated by the compressional loss K''. This assumption is supported by the T dependence of the Poisson ratio:

$$\sigma = \frac{0.5(u_L/u_T)^2 - 1}{(u_L/u_T)^2 - 1}. \tag{2}$$

Here, u_L and u_T are the longitudinal and transversal sound velocities, measured by BS from a low molecular weight PMPS [3]. The ratio of the relaxational parts, K_r and G_∞ of the compression and shear modulus, respectively, is then given by

$$\frac{K_r}{G_\infty} = \frac{2(\sigma-1)(R-1)}{\sigma R} - \frac{4}{3} \tag{3}$$

with $R = (u_{L,\infty}/u_{L,0})^2$ being the relaxation strength of the longitudinal modulus; $R \approx 2$ over the T range of the ultrasonic measurements. Thus, for the asymptotic value at high T $\sigma = 0.5$, M'' becomes equal to K''. With descending temper-

ature, σ decreases ($\sigma \approx 0.35$), and near T_g the values of G_∞ and K_r become comparable. Finally, the T dependence of the orientational relaxation times for PMPS display systematic deviations from a single VFTH behaviour (Fig. 2). Therefore, the dynamic glass transition phenomenon cannot be described by a single structural parameter theory over the whole T-range.

Acknowledgements

This work has been facilitated by the bilateral scientific agreement between Greece and Germany. I.A. and F.S. are grateful for the hospitality of the Research Center of Crete. F.S. acknowledges the financial support of the "Stiftervetband Deutschen Wissenschaft". We thank Drs. D. Boese and J. Kanetakis for providing the DS and FPI-DRS data.

References

1. Onabajo K, Dorfmüller Th, Fytas G (1987) J Polym Sci 25:749–763
2. Boese D, Momper B, Meier G, Kremer F, Hagenah JU, Fischer EW (1989) Macromolecules 22:4416–4421
3. Li BY, Wang CH (1989) J Chem Phys 90:2971–2978
4. Meier G, Fujara F, Petry W (1989) Macromolecules 22:4421–4425
5. Scilence SM, Goates SR, Nelson KH (1991) J Non-Cryst Solids 131:37–41
6. Alvarez F, Colmenero J, Fytas G (1992) (to be published)
7. Floudas G, Fytas G, Alig I (1991) Polymer 32:2307–2311
8. Angell CA (1991) J Non-Cryst Solids 131–133:13–31
9. Rössler E (1990) J Chem Phys 92:3725–3735
10. Floudas G, Higgins JS, Fytas G (1992) J Chem Phys 96:7672–7282

Authors' address:

Prof. G. Fytas
Foundation for Research and Technology-Hellas
P.O. Box 1527
711 10 Heraklion
Crete, Greece

Progress in Colloid & Polymer Science

Progr Colloid Polym Sci 91:165–170 (1993)

Diffusion in rod/sphere composite liquids

M. A. Tracy and R. Pecora

Department of Chemistry, Stanford University, Stanford, USA

Abstract: The synthesis, characterization, and some studies of the dynamics of a new rod/sphere composite liquid are discussed here. The composite liquid consists of silica spheres (39.4 and 60.4 nm radius) and poly(γ-benzyl-α,L-glutamate) rods (102 000 and 249 700 g/mol molecular weight) dispersed without aggregation in dimethylformamide (DMF) or a mixture of DMF and pyridine. The solvent mixture was used to refractive-index-match dilute solutions of spheres. Dynamic light scattering (DLS) was used to measure the diffusion constants of both the rods and the spheres simultaneously in the solution. Two dynamical regimes were observed for the diffusion of the rods with or without the spheres. The transition point between these two regimes was attributed to the change from dilute to semidilute dynamics. Sphere diffusion constants up to a value two times larger than expected from the Stokes–Einstein equation were measured for the composite liquid containing the large rod (249 700 g/mol) and the small sphere (39.4 nm) in accordance with the argument of Langevin and Rondelez [17]. The sphere diffusion constants measured for the liquid containing the large sphere (60.4 nm) and the small rod (102 000 g/mol) followed a stretched exponential decay with increasing polymer concentration: $D/D_0 = \exp(-0.16 c^{0.81})$. The $c^{0.81}$ dependence compares well to experimental results using coil polymers but not to the theories for rod solutions which predict a $c^{0.5}$ dependence. The similar concentration dependence found for rod and coil solutions suggests that polymer flexibility alone does not account for the deviation from the theoretical $c^{0.5}$ predictions. Instead, rod–rod hydrodynamic interactions may play an important role in this deviation.

Key words: Dynamic light scattering – rod/sphere composite liquids – diffusion

Introduction

Composite liquids – liquids composed of polymers, particles, and small molecule solvents – constitute an important class of synthetic and naturally-occurring materials. Examples include molecular composites, ceramic precursors, lubricants, paints, adhesives, and the cytoplasm in biological cells. Due to the complexity of these liquids, experimental studies of precisely defined systems are essential in developing an understanding of the interactions between all components in the liquid. Unfortunately, such fundamental studies have been relatively rare due to both the difficulty of synthesizing precisely defined composite liquids and the lack of adequate experimental methods to monitor the motions of the various constituents.

We have recently reported the synthesis, characterization and some studies of the dynamics of a rod/sphere composite liquid system [1]. In our case the "polymer" constituent is a rigid-rod polymer, poly(γ-benzyl-α,L-glutamate) (PBLG). Rigid-rod polymers are frequently used in composite liquids as viscosity enhancers. PBLG is commercially available in a wide range of molecular weights and its static and dynamic behavior have been characterized in dilute and semidilute solutions. It, in addition, forms mesophases in the concentrated regime. The "particles" in our composite liquid are coated silica spheres. These spheres are synthesized by the method of Stöber et al. [2] and coated with an organic coating (3-(trimethoxysilyl)propyl methacrylate (TPM)) following a procedure based on that of Philipse et al.

[3] to render them dispersible in organic solvents. The spheres with sizes in the range from 10 nm up to almost 1 μm can be synthesized with a relatively narrow size distribution. The solvent in our studies is dimethylformamide (DMF) and mixtures of DMF and pyridine. Both polymer and particle are dispersible as singlets (non-aggregating) and the PBLG retains its rigid (or nearly rigid) rod conformation in these solvents. The diffusion of both the polymer and the sphere in the composite liquid is measured by dynamic light scattering (DLS) [4].

Experiment

Coated silica spheres of radius 60.4 nm (determined by DLS) were synthesized by methods described previously [1]. The PBLG, purchased from Sigma, was fully characterized in DMF by total intensity light scattering (TILS). The molecular weight (M) was found to be 102 000 g/mol and the radius of gyration was 19 nm.

The dynamics of both binary rod/DMF and sphere/DMF solutions were studied by DLS in addition to the rod/sphere/DMF composite liquid. The rod concentrations studied varied from 1 to 30 mg/ml ($2 < nL^3 < 60$, where n is the rod number concentration and L is the rod length (70 nm)). The sphere concentrations were dilute (0.047–0.28 mg/ml). Sphere/DMF solutions were studied but showed no change in the diffusion constant over the dilute concentration range present in these experiments. DLS data were taken at a minimum of three of the following angles: 25.45° or 30.56°, 59.44°, 90°, and 110.35° or 120.56°. In addition, preliminary results are reported here for two solutions both consisting of PBLG of molecular weight 249 700 g/mol in the concentration range 1–8.3 mg/ml ($12 < nL^3 < 100$) but differing in the size of the spheres. In one case, spheres of radius 60.4 nm were used and in the other the radius was 39.4 nm. All experiments were done at 25 °C. Solution viscosities were measured with an Ostwald capillary viscometer at 25 °C.

The intensity autocorrelation function was measured in the homodyne DLS experiments using a standard DLS apparatus [1]. For the solutions described here, the diffusion constants of the rods and the spheres were determined from the experimentally determined correlation function using the FORTRAN program CONTIN which calculates

the distribution of decay times (τ_R) leading to the exponential decay of the measured correlation function. The decay time is inversely proportional to the diffusion constant, D:

$$\tau_R = 1/q^2 D , \qquad (1)$$

where q is the magnitude of the scattering vector,

$$q = (4\pi n/\lambda_0) \sin(\theta/2) , \qquad (2)$$

and n is the solvent index of refraction, λ_0 the wavelength of laser light in a vacuum, and θ the scattering angle. In the CONTIN results shown in Fig. 1, the distribution of decay times is presented as a plot of relative intensity versus hydrodynamic radius (R_h) of the scattering species. R_h is calculated from D using the Stokes–Einstein equation:

$$D = kT/6\pi\eta R_h , \qquad (3)$$

where k is the Boltzmann constant, T the absolute temperature, and η the solution viscosity. The solvent viscosity was used in place of the solution viscosity in the outputs to provide a standard for comparison of all peaks. The intensity at a given R_h is the fraction of scattered light from particles of "size" R_h.

Results

As shown in Fig. 1, both the rod and the sphere diffusion constants were measured simultaneously in the composite liquid solutions. Figure 1 also shows that the peaks in the composite liquid DLS results were unambiguously identified from a comparison with the binary sphere/DMF and rod/DMF DLS results. In all the composite liquid data analyzed two dominant peaks were resolved by CONTIN. The slow peak was attributed to the translational diffusion of the spheres because as the polymer concentration decreased, the hydrodynamic radius approached that of the spheres alone in DMF. The faster peak (smaller R_h) is due to the translational diffusion of the rods. The variation of the rod translational diffusion constant with polymer concentration shown in Fig. 2 closely follows that of the binary system confirming that the peak does belong to the rods. More importantly, this indicates that the spheres do not affect rod diffusion at these low sphere concentrations. Both the sphere and rod diffusion constants are

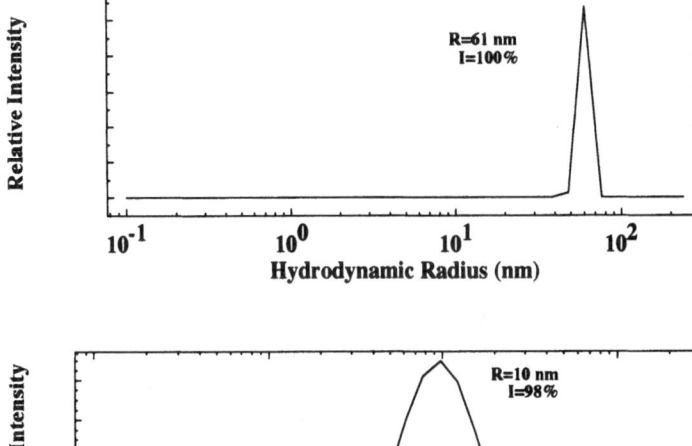

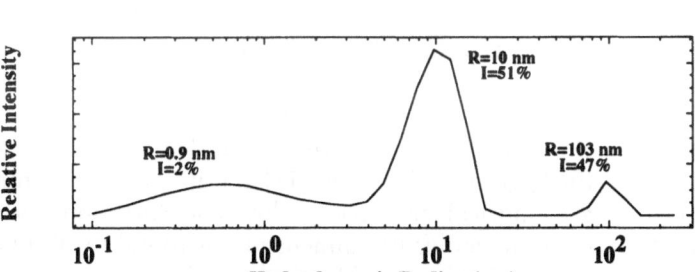

Fig. 1. Comparison of binary solution DLS results to those of the ternary composite liquid. The top figure is a typical intensity vs. hydrodynamic radius plot obtained from CONTIN for a 0.01% silica/DMF solution. Each CONTIN peak is marked by an R and an I. R is the average hydrodynamic radius of the peak and I is the fraction of the total scattered intensity within that peak. The second plot is a typical CONTIN output for a 0.5% PBLG/DMF solution. The bottom plot is a typical output for the composite liquid showing both the rod and the sphere diffusion. This composite liquid contained 0.5% rods and 0.005% spheres in DMF. For each output, the solvent viscosity, not the solution viscosity, was used to calculate R. Thus, R for the sphere varies from the value in the sphere/DMF output. These examples were taken from data measured at 90°

independent of angle. The very fast, weak third peak is most likely due to the translational–rotational coupling of the rod diffusion [5].

Another important characteristic of this composite liquid system is that the sphere's contribution to the total light scattered can be reduced or eliminated using a DMF/pyridine solvent mixture. The indices of refraction of DMF and pyridine at 25 °C at the sodium D line (n_D) are 1.42817 and 1.50745, respectively. At a solvent composition of 31.8% pyridine/68.2% DMF, it was found that the sphere peak disappeared from the CONTIN results for a solution containing 1.0 mg/ml rods and 0.047 mg/ml silica. At this point, the sphere index of refraction was calculated to be 1.4528 by the Dale–Gladstone equation [6]. The PBLG refractive index is 1.59 [7]. The sphere peak went from contributing as much as 80% of the scattered light

intensity at 0% pyridine to 0% of the intensity at 31.8% pyridine. Thus, refractive index matching by varying the solvent composition allows solutions with a high sphere concentration to be studied using optical techniques without complications due to multiple scattering and turbidity.

Discussion

Several comments concerning the dynamics of both binary rod/solvent and ternary rod/sphere/solvent solutions may be made from the results of these experiments. As shown in Fig. 2, two dynamical regimes were observed for the rods. The average diffusion constants shown here were determined at each rod concentration from the slope of a linear least-squares fit to a plot of $1/\tau_R$ vs. q^2. The diffusion constant errors were calculated

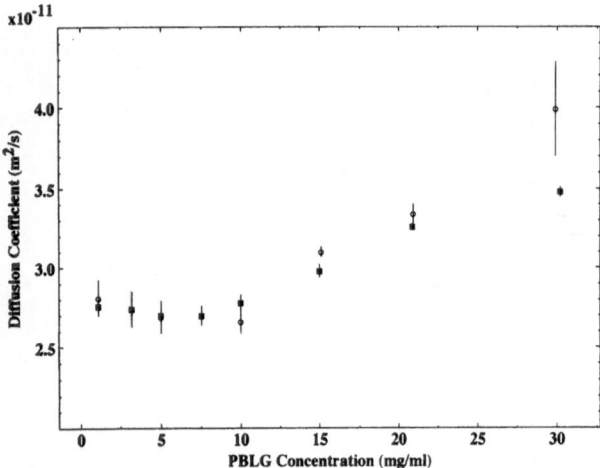

Fig. 2. Rod translational diffusion coefficient versus concentration. The rod translational diffusion coefficients from the CONTIN results for the binary rod/DMF solutions (■) are compared to the rod diffusion coefficients from the CONTIN results for the composite liquid (○) at similar rod concentrations. The sphere concentrations in the composite liquids ranged from 0.04 mg/ml to 0.28 mg/ml SiO_2

from the errors in the linear fit. For solutions of coil polymers, two dynamical regimes have also been found [8–10] and have been attributed to a transition between dilute and semidilute dynamics. Brown and Mortensen [8] noted that this transition occurs $C^* \approx 1/[\eta]$, where $[\eta]$ is the intrinsic viscosity. From PBLG viscosity data taken for this sample, $C^* = 8.4$ mg/ml ($nL^3 = 17$). This C^* is very near where the transition occurs in Fig. 2. This agrees well with depolarized light scattering results on PBLG [5].

From the dilute regime data, PBLG was found to behave as a rigid hard rod in DMF with a length of 70 nm, a geometric diameter of 1.63 nm, and a hydrodynamic diameter (d) of 2.08 nm [1]. Extrapolating the dilute regime data to zero concentration gives an infinite dilution diffusion constant (D_0) of $2.77 \cdot 10^{-11}$ m²/s. The Broersma equation for D_0 of a rigid rod of $L/d > 5$ was used to calculate the hydrodynamic diameter. The calculated d was in the range expected for PBLG from previous experiments. Models for short rods and worm-like chains predicted diameters higher than the values found experimentally [1]. The geometric diameter was calculated from the second virial coefficient (A_2) measured by TILS using theories for hard rods.

In the semidilute regime there is a modest increase in the diffusion constant with increasing concentration (c). Slopes were calculated from linear fits to the dilute and semidilute data. From the slopes and intercepts, values of k_D were determined according to

$$D = D_0(1 + k_D c) . \tag{4}$$

The diffusion virial coefficient, k_D, is given by

$$k_D = 2MA_2 - k_f - \nu , \tag{5}$$

where ν is the specific molar volume (0.791 cm³/g for PBLG) and k_f is the coefficient in the linear term in the expansion of the friction factor in powers of the concentration. As may be seen from the above equations, k_D is a measure of the competition between thermodynamic effects which, in the case of PBLG, speed up diffusion and frictional effects which slow it down [11]. Since Eq. (4) is valid at low concentrations, where only pair correlations are significant, it is not surprising that this equation does not account for the change from negative to positive k_D seen here. Equation (4) was used to calculate the experimental k_f values for the different concentration regimes. The theoretical values for k_f given by Peterson [12] and Itou et al. [13] for a cylindrical hard rod at low concentration with the same molecular dimensions as found for PBLG of $L = 70$ nm are identical to the experimental k_f values within error. This further suggests that the PBLG behaves as a stiff hard rod in solution. The strictly repulsive hard rod potential appears to be sufficient to describe the diffusion data in the dilute regime. The theories of Doi et al. [14–16] for the DLS dynamic structure factor of semidilute and concentrated solutions of long rigid rods give neither qualitative nor quantitative descriptions of our DLS data even though the concentrations studied should fall within the range of their theory for semidilute solutions [1].

In the ternary system, the sphere diffusion constant was found to follow the Stokes–Einstein equation for the composite liquid consisting of spheres of $R = 60.4$ nm in a solution of rods of $M = 102\,000$ g/mol (composite liquid 1) at all polymer concentrations studied. Preliminary results for a second composite liquid (composite liquid 2), consisting of the same size spheres in a solution of rods of $M = 249\,700$ g/mol indicate that, at the polymer concentrations studied, the Stokes–

Einstein relation also holds up to a concentration of 8.3 mg/ml. A third composite liquid (composite liquid 3) was synthesized in which the sphere radius was reduced to $R = 39.4$ nm. The diffusion constant of these smaller spheres was measured by DLS in a solution of rods of $M = 249\,700$ g/mol in the same rod concentration range used in the composite liquid 2 studies. Unlike composite liquids 1 and 2, sphere diffusion coefficients up to a value two times larger than expected from the Stokes–Einstein equation using the solution viscosity were measured for composite liquid 3. The maximum deviation occurred at about C^*.

This behavior can be explained if the polymer solution structure resembles a transient "net" above C^* as suggested by Langevin and Rondelez [17]. According to their argument, the Stokes–Einstein relation holds as the spheres see the macroscopic polymer solution viscosity. But as the sphere size becomes less than the net hole size above C^*, the spheres are able to diffuse through these holes and, thus, see an effective viscosity that approaches that of the pure solvent. As the polymer concentration increases, the hole size in this transient rod "net" decreases so the Stokes–Einstein behavior is again approached.

Finally, the sphere diffusion in composite liquid 1 was found to follow the stretched exponential relation:

$$D/D_0 = \exp(-\alpha c^\nu), \qquad (6)$$

where $\alpha = 0.16 \pm 0.02$, $\nu = 0.81 \pm 0.07$, and c is the polymer concentration (in mg/ml). The values for α and ν were determined from a stretched exponential fit to the sphere diffusion constant data for composite liquid 1 at all polymer concentrations (1–30 mg/ml) [1]. The value for ν compares well to the empirical prediction of Phillies [18] and the experimental values determined by Zhou and Brown [19] for silica/polyisobutylene coils. However, these results are in between $\nu = 0.91$ found for silica/polymethyl methacrylate [20] and $\nu = 0.62$ found for silica/polyethylene oxide [17]. Theories predicting such a stretched exponential relationship for spheres diffusing through rigid rods all predict a $c^{0.5}$ dependence. Altenberger et al. [21] predict an exponent varying between 0.5 and 1.0, depending on the hydrodynamic interactions between points in the polymer network and the concentration of the obstacles. At high concentrations or strong hy-

drodynamic interactions, the c dependence is stronger than $c^{0.5}$. Thus, the theories are not able to exactly predict the observed value for ν. Since the deviation from $c^{0.5}$ behavior was found for solutions of rods and coils, this discrepancy is not due to polymer flexibility alone. The theory of Altenberger et al. [21] suggests, however, that hydrodynamic interactions between the rods are important in characterizing sphere transport through a rod network. There is even less agreement on the theoretical significance of α which is often thought to depend in part on the sphere radius. Current models are evidently inadequate to explain the diffusion of probe spheres in polymer/solvent systems.

Conclusion

We have synthesized a new rod/sphere composite liquid and performed DLS experiments to study the rod and sphere dynamics at a variety of concentrations. The diffusion constants of both the rods and the spheres were measured simultaneously using DLS. The binary PBLG/DMF results revealed two distinct dynamical regimes which were identified as the dilute and semidilute rod concentration regimes. The TILS and DLS data suggest that in DMF, PBLG is best modeled as a long rigid hard rod. In the ternary system, deviations from the Stokes–Einstein equation were observed. Deviations could be qualitatively explained if the polymer solution is viewed as a "net" with holes rather than as a continuum. Finally, sphere diffusion was found to follow a stretched exponential relationship between the sphere diffusion constant and rod concentration. Though the concentration exponent differs from the theoretical dependence, it is similar to that found in silica/coil polymer systems. Thus, polymer flexibility alone does not adequately explain the difference between experiment and theory.

References

1. Tracy MA, Pecora R (1992) Macromolecules 25:337
2. Stöber W, Fink A, Bohn E (1968) J Colloid Interface Sci 26:62
3. Philipse A, Vrij A (1989) J Colloid Interface Sci 128:121
4. Berne BJ, Pecora R (1976) Dynamic Light Scattering. Wiley, New York
5. Zero KM, Pecora R (1982) Macromolecules 15:87

6. Huglin MB (ed.) (1972) Light Scattering from Polymer Solutions. Academic Press, London
7. Block H (1983) Poly(γ-benzyl-L-glutamate) and Other Glutamic Acid Containing Polymers. Ch. 5 Gordon and Breach, New York
8. Brown W, Mortensen K (1988) Macromolecules 21:420
9. Adam M, Delsanti M (1977) Macromolecules 10:1229
10. Mathiez P, Mouttet C, Weisbuch G (1981) Biopolymers 20:2381
11. Russo PS, Karasz FE, Langley KH (1984) J Chem Phys 80:5312
12. Peterson J (1964) J Chem Phys 40:2680
13. Itou S, Nishioka N, Norisuye T, Teramoto A (1981) Macromolecules 14:904
14. Shimada T, Doi M, Okano K (1988) J Chem Phys 88:2815
15. Doi M, Shimada T, Okano K (1988) J Chem Phys 88:4070
16. Shimada T, Doi M, Okano K (1988) J Chem Phys 88:7181
17. Langevin D, Rondelez F (1978) Polymer 19:875
18. Phillies GDJ (1989) J Phys Chem 93:5029
19. Zhou P, Brown W (1989) Macromolecules 22:890
20. Brown W, Rymdén R (1988) Macromolecules 21:840
21. Altenberger AR, Tirrell M, Dahler JS (1986) J Chem Phys 84:5122

Received December 16, 1991;
accepted June 15, 1992

Authors' address:

Mark A. Tracy
Enzytech Inc.
64 Sidney St
Cambridge, MA 02139, USA

Author Index

Alegría A   24
Alig I   162
Alvarez F   20, 24
Anastasiadis SH   88, 97
Ansarifar MA   83
Arbe A   24

Bahar I   1, 16
Baschnagel J   5
Bastide J   105
Benmansour Z   109
Benmouna M   109
Benoit H   109
Berman R   43
Binder K   5
Binkert Th   156
Börjesson L   43, 46
Boué F   105
Brereton MG   8
Brown W   113
Burgess A   28
Buzier M   105

Chu B   51, 142
Colmenero J   20, 24

Dai L   83
Darinskii A   13
Duval M   117

Erman B   1, 16
Ewen B   121, 130

Farago B   121, 130
Fetters LJ   130
Fischer EW   1, 35, 58, 66, 109
Fleischer G   55
Floudas G   28, 124, 162
Frick B   24
Fytas G   20, 58, 72, 124, 135, 162

Gallot Y   117
Gerharz B   58
Giebel L   124
Gießler KH   101
Gläser H   35

Gotlib Yu   13
Götzelmann A   101

Haida H   117
Heermann DW   5
Hellmann EH   146
Hellmann GP   146
Helmstedt M   127
Higgins JS   28
Hoffmann A   61
Huang JS   130

Jacobsson P   46
Janßen S   80
Joosten JGH   149

Karasz FE   153
Koch T   61
Kremer F   1, 39
Kremer K   5

Langley KH   153
Lartigue C   105
Li Y   51
Lindner P   105
Lingelser JP   117
Lohfink M   31
Lukyanov M   13
Lyulin A   13

Majkrzak CF   88, 97
Maschke U   121
Meewes M   156
Meier G   35, 66
Mendes E   105
Menelle A   88, 97
Momper B   66
Mortensen K   69
Motschmann H   83

Neelov I   13
Ngai KL   72, 75, 135
Nilgens H   35

Oeser R   105

Patkowski A   35
Paul W   5
Pecora R   165

Quellet Ch   156

Rauch F   101
Reiter G   93
Rennie AR   146
Richter D   121, 130
Ricka J   156
Rizos AK   72, 135
Roland CM   75
Roovers JEL   72
Russell TP   88, 97

Satija SK   88, 97
Schlosser E   39, 158
Schönhals A   39, 158
Schuler M   61
Schwahn D   80
Sidebottom DL   43, 46
Sillescu H   31
Springer T   80
Stamm M   83, 101
Steffen W   35, 124
Steiner U   93
Stickel F   61
Stieber F   162
Stühn B   61

Teraoka I   153
Toprakcioglu C   83
Torell LM   43, 46
Tracy MA   165

Vilgis TA   109
Vogt S   58

Wang CH   20, 138
Wang Z   142
Wittmann H-P   5

Yu J   142

Zielinski F   105

Subject Index

α-relaxation   39, 158
–, dynamics of the   24

adsorption   83
anisotropic polarizability, dense
    systems   16
annealing effects   51
apparatus, surface-force   83
Arrhenius plot   158

blends, polymer   8, 146
block copolymers   61, 69, 83, 97
bond fluctuation model   5
branching   127
bulk state, *cis*-polyisoprene in the   1

chain diffusion   146
– dynamics, local   1, 16
– fragmentation   146
–, freely rotating   16
chains with fixed ends   16
characterization, polymer   127
cis-polyisoprene in the bulk state   1
cluster size, equilibrium   35
clusters   105
coefficient, Onsager   66
coil-globule-transition   142, 156
colloidal crystal   69
composite liquids, rod/sphere   165
composition fluctuations   58
computer simulation   13
concentrated solutions   113, 124
concentration fluctuations   72, 75
configurational transitions   16
cooperative diffusion   138
– mode   109
copolymer, diblock   58, 72, 88, 101,
    117
–, block   61, 69, 83, 97
correlation length, static, dynamic   35
– spectroscopy, photon   61
–, photon   113
coupling model   72, 135
–, mode-mode   66
critical phenomena   66
crystal, colloidal   69
cubatic   69

decomposition, spinodal   80
dense systems anisotropic
    polarizability   16
depolarized Rayleigh scattering   58
di-2-ethylhexyl phtalate   28

diblock copolymer   58, 72, 88, 101,
    117
dielectric measurements   158
– relaxation   39, 124
– spectroscopy 20, 58
diffusion   31, 153, 165
–, chain   146
–, cooperative   138
–, segmental   121
–, self-   55
–, tracer   55, 149
dynamic correlation length, static   35
– light scattering   72, 117, 135, 138,
    153, 165
– – –, static   35
–, local chain   1, 16
–, microscopic   130
– of glass forming liquids   35
– of the α-relaxation   24
–, polymer   127
– rotational isomeric state
    formalism   1
– scattering   109

effects, annealing   51
ellipsometry   83
ends, chains with fixed   16
entanglements   130, 158
environment, fluctuating   1
–, restrictive   16
equilibrium cluster size   35

films, thin   88, 97, 101
fixed ends, chains with   16
fluctuating environment   1
fluctuation model, bond   5
–, composition   58
–, concentration   72
formalism, dynamic rotational isomeric
    state   1
forming liquids, dynamics of glass   35
fragmentation, chain   146
freely rotating chain   16

gels   105
glass forming liquids, dynamics of   35
– – polymeric systems   24
–, porous   153
– transition   5, 8, 24, 28, 31, 43

incoherent neutron scattering   28
interactions, polymer-surfactant   156
interdiffusion   8

interdiffusive mode   109
isomeric state formalism, dynamic
    rotational   1

kinetics, phase separation   51

laser light scattering   142
length, static, dynamic correlation   35
light scattering   43, 46, 66, 127, 149
– –, dynamic 72, 117, 135, 138, 153,
    165
– –, laser   142
– –, static, dynamic   35
liquids, dynamics of glass forming   35
–, rod/sphere composite   165
local chain dynamics   1, 16
longitudinal relaxation   162

material, viscoelastic   80
measurements, dielectric   158
media, non-ergodic   149
melts   105, 121
–, polymer   130
method, Monte Carlo   5
micelle   69
microphase separation   51
– – transition   61
microscopic dynamics   130
miscible blend   75
mixtures, polymer   66, 121
–, ternary   109
mobility   13
mode-mode coupling   66
model, bond fluctuation   5
–, coupling   72, 135
modification, solvent   135
Monte Carlo method   5
motion, segmental   1, 80

networks   121
neutron reflectivity   88, 93, 97
– reflectometry   83, 101
– scattering   105, 146
– –, incoherent   28
– –, quasielastic   24
– spin-echo   130
– – spectroscopy   121
NMR, PFG-   55
non-ergodic media   149
nonexponential behavior   39
nonselective solvent   117
normal-mode relaxation   158
nuclear reaction analysis   93, 101

Onsager coefficient 66
order 101
ordering, surface 97

PCS 20
percolation 105
PFG-NMR 55
phase separation kinetics 51
phenomena, critical 66
photon correlation 113
– – spectroscopy 61, 124
phtalate, di-2-ethylhexyl 28
PIMA 20
plasticizers 31
plot, Arrhenius 158
PMMA 20
polarizability, dense systems
    anisotropic 16
poly(ethylene oxide) 55
poly(methylphenyl siloxane) 162
poly(N-isopropylacrylamide) 156
poly(siloxane) 55
polyelectrolyte 142
polyethylene 127
polymer 31, 105
– blends 8, 80, 146
– chain collapse 142
– characterization 127
– diffusion 93
– dynamics 127
–, LC 13
– melts 130
– mixtures 66, 121
–, semirigid 153
– solution 138
polymer-surfactant interactions 156
polymeric systems, glass-forming 24
polystyrene 113, 142
polyurethanes, segmented 51
porous glass 153
primary relaxation 58

quasielastic neutron scattering 24

Rayleigh scattering, depolarized 58
reflection, neutron 93
reflectivity, neutron 88, 97
reflectometry, neutron 83, 101
relaxation 20, 43, 80
–, dielectric 124
–, longitudinal 162
–, normal-mode 158
–, primary 58
–, reorientational 162
–, structural 46
– techniques 24
– times 113
reorientational relaxation 162
reptation 93, 130
restrictive environment 16
rod/sphere composite liquids 165
rotating chain, freely 16
rotational isomeric state formalism,
    dynamic 1
Rouse model 93

scaling behavior 43
scattering, dynamic 109
–, dynamic light 72, 117, 135, 138,
    153, 165
–, incoherent neutron 28
–, laser light 142
–, light 43, 46, 66, 127, 149
–, neutron 105, 146
–, quasielastic neutron 24
–, small-angle x-ray 61
–, static, dynamic light 35
segmental diffusion 121
– motion 1, 80
– relaxation 75
segmented polyurethanes 51
self-diffusion 5, 55
semidilute solution 117, 127
semirigid polymers 153
separation kinetics, phase 51
–, microphase 51

– transition, microphase 61
simulation, computer 13
size, equilibrium cluster 35
small-angle x-ray scattering 61
solution, concentrated 113, 124
–, polymer 138
–, semidilute 117, 127
solvent modification 135
–, nonselective 117
spectroscopy, dielectric 20, 58
–, neutron spin echo 121
–, photon correlation 61, 124
sphere composite liquids, rod 165
spin-echo, neutron 130
spinodal decomposition 80
state formalism, dynamic rotational
    isomeric 1
–, *cis*-polyisoprene in the bulk 1
static, dynamic correlation length 35
–, dynamic light scattering 35
strong-fragile 46
structural relaxation 46
surface ordering 97
surface-force apparatus 83
synchrotron SAXS 51
systems anisotropic polarizability,
    dense 16
–, glass-forming polymeric 24

techniques, relaxation 24
ternary mixtures 109
times, relaxation 113
tracer diffusion 55, 149
transition, coil-to-globule 142, 156
–, configurational 16
–, glass 5, 8, 24, 28, 31, 43
–, microphase separation 61

viscoelastic material 80
viscoelasticity 138

x-ray scattering, small-angle 61